Anatomy & Physiology

made
Incredibly
Easy!®

Fifth Edition

D1214846

Anatomy & Physiology

made Incredibly Easy!®

Fifth Edition

Clinical Editor
Laura M. Willis, DNP, APRN, FNP-C
Family Nurse Practitioner
Urbana Family Medicine and Pediatrics
Urbana, Ohio

Wolters Kluwer

Philadelphia • Baltimore • New York • London
Buenos Aires • Hong Kong • Sydney • Tokyo

Executive Editor: Nicole Dernoski
Development Editor: Maria M. McAvey
Editorial Coordinator: Amberly Hyden
Production Project Manager: Marian Bellus
Design Coordinator: Elaine Kasmer
Manufacturing Coordinator: Kathleen Brown
Marketing Manager: Lindsay Dierse
Prepress Vendor: SPi Global

5th edition

Copyright © 2018 Wolters Kluwer

9 8 7 6 5 4 3 2

Printed in China (or the United States of America)

Library of Congress Cataloging-in-Publication Data
Names: Willis, Laura M., 1969- editor.
Title: Anatomy & physiology made incredibly easy! / clinical editor, Laura M. Willis.
Other titles: Anatomy and physiology made incredibly easy!
Description: Fifth edition. | Philadelphia : Wolters Kluwer, [2018] | Includes bibliographical references and index.
Identifiers: LCCN 2017017763 | ISBN 9781496359162
Subjects: | MESH: Anatomy | Physiological Phenomena
Classification: LCC QP41 | NLM QS 4 | DDC 612—dc23 LC record available at https://lccn.loc.gov/2017017763

LWW.com

Dedication

Dedicated to my family: without you I could not have made it to where I am and I am so grateful. You are my joy!

Laura M. Willis

Contributors

Melanie N. DeGonzague, MSN, APRN, AGPCNP-BC, CNP
Internal Medicine, The Christ Hospital
 Medical Associates
Cincinnati, Ohio

Kathryn Dinh, MSN, APRN, AGPCNP-BC
Nurse Practitioner
The Christ Hospital Diabetes & Endocrine
 Center
Cincinnati, Ohio

Shelba Durston, MSN, RN, CCRN, SAFE
Professor of Nursing
San Joaquin Delta College
Stockton, CA
Staff Nurse IV, ICU/CCU/Sexual Assault
 Forensic Examiner
San Joaquin General Hospital
French Camp, CA

Kay L. Luft, MN, CCRN, CNE
Associate Professor
Saint Luke's College of Health Sciences
Kansas City, MO

Katrin Moskowitz, DNP, FNP
Family Nurse Practitioner
CHC, Inc.
Meriden, Connecticut

Deborah Skoruppa, DNP, APRN, NP-C, CNE
Professor of Nurse Education
Nurse Education, Del Mar College
Corpus Christi, Texas

Tracy Taylor, MSN, RN
Adjunct Faculty
Ohio Institute of Allied Health
Dayton, Ohio
NICU Nurse
Miami Valley Hospital
Dayton, Ohio

Stefanie Tyler, DNP, WHNP
Royal Bournemouth Hospital
NHS Foundation Trust
Bournemouth, Dorset, UK

Previous Edition Contributors

Dometrives Armstrong, MSN, FNP, PHN

Cheryl L. Brady, MSN, RN

Shelba Durston, MSN, RN, CCRN

Ruth Howell, BSN, MEd

Karla R. Jones, MS, RN

Kay Luft, MN, CCRN, CNE

Betty Sims, MSN, RN, FRE

Deborah Skoruppa, MSN, RN, NP-C, CNE

Cynthia Small, MSN, FNP-BC, CNE

Brigitte Thiele, BSN, RN

Peggy Thweatt, MSN, RN

Foreword

If you're like me, you're too busy to wade through a foreword that uses pretentious terms and umpteen dull paragraphs to get to the point. So let's cut right to the chase! Here's why this book is so terrific:

1. It will teach you all the important things you need to know about anatomy and physiology without tons of tiny details to wade through.
2. It will help you remember what you've learned.
3. It will make you smile as it enhances your knowledge and skills.
   Don't believe me? Try these recurring logos on for size:

 Zoom in—provides a close look at anatomic structures

 Body shop—helps explain how body systems and structures work together

 Now I get it!—converts complex physiology into easy-to-digest explanations

 Senior moment—pinpoints the effects of aging on anatomy and physiology

 Memory jogger—reinforces learning through acronyms and other tools that aid recall

 Just for fun—reinforces understanding of anatomical terminology and pathophysiological concepts

See? I told you! And that's not all. Look for me and my friends in the margins throughout this book. We'll be there to explain key concepts, provide important care reminders, and offer reassurance. Oh, and if you don't mind, we'll be spicing up the pages with a bit of humor along the way, to teach and entertain in a way that no other resource can.

I hope you find this book helpful. Best of luck throughout your career!

Contents

Chapter 1

The human body

Just the facts

In this chapter, you'll learn:

♦ anatomic terms for direction, reference planes, body cavities, and body regions to help describe the locations of various body structures
♦ the structure of cells
♦ cell reproduction and energy generation
♦ four basic tissue types and their characteristics.

Locating body structures starts with directional terms, reference planes, cavities, and regions.

Anatomic terms

Anatomic terms describe directions within the body as well as the body's reference planes, cavities, and regions.

Directional terms

When navigating the body, directional terms help you determine the exact location of a structure.

Couples at odds

Generally, directional terms can be grouped in pairs of opposites:
- *Superior* and *inferior* mean above and below, respectively. For example, the shoulder is superior to the elbow, and the hand is inferior to the wrist.
- *Anterior* means toward the front of the body, and *posterior* means toward the back. *Ventral* is sometimes used instead of anterior, and *dorsal* is sometimes used instead of posterior.
- *Medial* means toward the body's midline and *lateral* means away from it.
- *Proximal* and *distal* mean closest and farthest, respectively, to the point of origin (or to the trunk).
- *Superficial* and *deep* mean toward or at the body surface and farthest from it.

Memory jogger

To remember the meanings of **proximal** and **distal**, keep in mind that when something is in **proximity**, it's nearby. When something is **distant**, it's far away.

1

Reference planes

Reference planes are imaginary lines used to section the body and its organs. These lines run longitudinally, horizontally, and angularly.

The four major body reference planes are:
1. median sagittal
2. frontal
3. transverse
4. oblique. (See *Picturing body reference planes.*)

Body shop

Picturing body reference planes

Body reference planes are used to indicate the locations of body structures. Shown here are the median sagittal, frontal, and transverse planes. An oblique plane—a slanted plane that lies between a horizontal plane and a vertical plane—isn't shown.

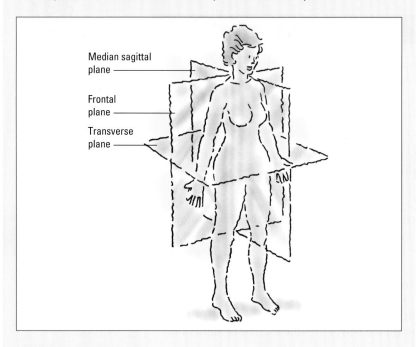

Median sagittal plane

Frontal plane

Transverse plane

Body cavities

Body cavities are spaces within the body that contain the internal organs. The *dorsal* and *ventral cavities* are the two major closed cavities—cavities without direct openings to the outside of the body. (See *Locating body cavities*.)

Dorsal cavity

The dorsal cavity is located in the posterior region of the body.

The think tank and backbone of the operation

The dorsal cavity is further subdivided into two cavities:
- The *cranial cavity* (also called the *calvaria*), formed by the skull, encases the brain.
- The *vertebral cavity* (also called the *spinal cavity* or *vertebral canal*), formed by the vertebrae, encloses the spinal cord.

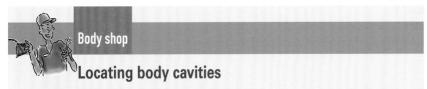

Body shop

Locating body cavities

The dorsal cavity, in the posterior region of the body, is divided into the cranial and vertebral cavities. The ventral cavity, in the anterior region, is divided into the thoracic and abdominopelvic cavities. These regions are shown in the illustration given below.

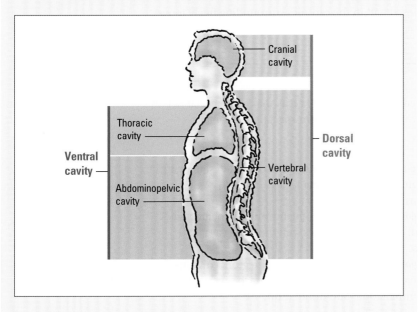

Ventral cavity

The ventral cavity occupies the anterior region of the trunk. This cavity is subdivided into the *thoracic cavity* and the *abdominopelvic cavity*.

Treasure chest

Surrounded by the ribs and chest muscles, the thoracic cavity refers to the space located superior to the abdominopelvic cavity. It's subdivided into the *pleural cavities* and the *mediastinum*:
- Each of the two *pleural cavities* contains a lung.
- The *mediastinum* houses the heart, large vessels of the heart, trachea, esophagus, thymus, lymph nodes, and other blood vessels and nerves.

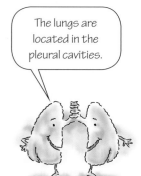

The lungs are located in the pleural cavities.

The bread basket and below

The abdominopelvic cavity has two regions, the *abdominal cavity* and the *pelvic cavity:*
- The *abdominal cavity* contains the stomach, intestines, spleen, liver, and other organs.
- The *pelvic cavity*, which lies inferior to the abdominal cavity, contains the bladder, some of the reproductive organs, and the rectum.

Other cavities

The body also contains an *oral cavity* (the mouth), a *nasal cavity* (located in the nose), *orbital cavities* (which house the eyes), *middle ear cavities* (which contain the small bones of the middle ear), and the *synovial cavities* (enclosed within the capsules surrounding freely moveable joints).

Body regions

Body regions are used to designate body areas that have special nerves or vascular supplies or those that perform special functions.

The guts of the matter

The most widely used body region terms are those that designate the sections of the abdomen. (See *Abdominal regions exposed.*)

The abdomen has nine regions:
- The *umbilical region*, the area around the umbilicus, includes sections of the small and large intestines, inferior vena cava, and abdominal aorta.
- The *epigastric region*, superior to the umbilical region, contains most of the pancreas and portions of the stomach, liver, inferior vena cava, abdominal aorta, and duodenum.

- The *hypogastric region* (or pubic area), inferior to the umbilical region, houses a portion of the sigmoid colon, the urinary bladder and ureters, the uterus and ovaries (in females), and portions of the small intestine.
- The right and left *iliac regions* (or inguinal regions) are situated on either side of the hypogastric region. They include portions of the small and large intestines.
- The right and left *lumbar regions* (or loin regions) are located on either side of the umbilical region. They include portions of the small and large intestines and portions of the kidneys.
- The right and left *hypochondriac regions,* which reside on either side of the epigastric region, contain the diaphragm, portions of the kidneys, the right side of the liver, the spleen, and part of the pancreas.

Remember, each region has a specific nerve or vascular supply or performs a special function.

Body shop

Abdominal regions exposed

Here's an anterior view of the abdominal regions.

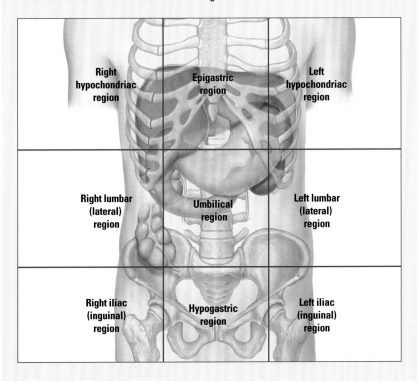

Right hypochondriac region

Epigastric region

Left hypochondriac region

Right lumbar (lateral) region

Umbilical region

Left lumbar (lateral) region

Right iliac (inguinal) region

Hypogastric region

Left iliac (inguinal) region

A look at the cell

Sometimes simpler is better. The simpler I am, the more I can regenerate!

The *cell* makes up the body's structure and serves as the basic unit of living matter. Human cells vary widely, ranging from the simple squamous epithelial cell to the highly specialized neuron.

The greatest regeneration

Generally, the simpler the cell, the greater its power to regenerate. The more specialized the cell, the weaker its regenerative power. Cells with greater regenerative power have shorter life spans than do those with less regenerative power.

Cell structure

Cells are made up of three basic components:
1. protoplasm
2. plasma membrane
3. nucleus. (See *Inside the cell.*)

Protoplasm

Protoplasm, a viscous, translucent, watery material, is the primary component of plant and animal cells. It contains a large percentage of water, inorganic ions (such as potassium, calcium, magnesium, and sodium), and naturally occurring organic compounds (such as proteins, lipids, and carbohydrates).

Getting charged

The inorganic ions within protoplasm are called *electrolytes.* They regulate acid-base balance and control the amount of intracellular water. When these ions lose electrons (minute particles with a negative charge), they acquire a positive electrical charge. When they gain electrons, they acquire a negative electrical charge. The most common electrolytes in the body are sodium (Na^+), potassium (K^+), and chloride (Cl^-)

I'm more than just a pretty face!

A pair of "plasms"

Nucleoplasm is the protoplasm of the cell's nucleus. It plays a part in reproduction. *Cytoplasm* is the protoplasm of the cell body that surrounds the nucleus. It converts raw materials to energy. It's also the site of most synthesizing activities. In the cytoplasm, you'll find *cytosol, organelles,* and *inclusions.*

Inside the cell

This cross section shows the components and structures of a cell. As noted, each component plays a part in maintaining the health of the cell.

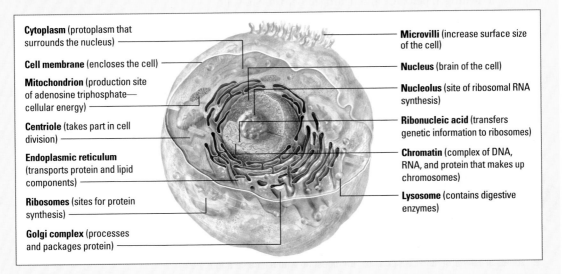

Cytoplasm (protoplasm that surrounds the nucleus)

Cell membrane (encloses the cell)

Mitochondrion (production site of adenosine triphosphate—cellular energy)

Centriole (takes part in cell division)

Endoplasmic reticulum (transports protein and lipid components)

Ribosomes (sites for protein synthesis)

Golgi complex (processes and packages protein)

Microvilli (increase surface size of the cell)

Nucleus (brain of the cell)

Nucleolus (site of ribosomal RNA synthesis)

Ribonucleic acid (transfers genetic information to ribosomes)

Chromatin (complex of DNA, RNA, and protein that makes up chromosomes)

Lysosome (contains digestive enzymes)

A cytosol sea

Cytosol is a viscous, semitransparent fluid that's 70% to 90% water. It contains proteins, salts, and sugars.

A lot to metabolize

Organelles are the cell's metabolic units. Each organelle performs a specific function to maintain the life of the cell:
- *Mitochondria* are structures within the cytoplasm that provide most of the body's adenosine triphosphate—the enzyme that fuels many cellular activities.
- *Ribosomes* are the sites of protein synthesis.
- The *endoplasmic reticulum* is an extensive network of membrane-enclosed tubules. *Rough endoplasmic reticulum* is covered with ribosomes and produces certain proteins. *Smooth endoplasmic reticulum* contains enzymes that synthesize lipids.

Organelles are my metabolic units.

- Each *Golgi apparatus* synthesizes carbohydrate molecules. These molecules combine with the proteins produced by rough endoplasmic reticulum to form secretory products such as lipoproteins.
- *Lysosomes* are digestive bodies that break down foreign or damaged material in cells. (See *Lysosomes at work.*)
- *Peroxisomes* contain *oxidases,* enzymes capable of reducing oxygen to hydrogen peroxide and hydrogen peroxide to water.
- *Cytoskeletal elements* form a network of protein structures.
- *Centrosomes* contain *centrioles,* short cylinders that are adjacent to the nucleus and take part in cell division.

Temps that don't do any work

Inclusions are nonfunctioning units in the cytoplasm that are commonly temporary. The pigment *melanin* in epithelial cells and the stored nutrient *glycogen* in liver cells are both examples of nonfunctioning units.

Now I get it!

Lysosomes at work

Lysosomes are the organelles responsible for digestion within a cell. Phagocytes assist in this process. Here's how lysosomes work.

Function of lysosomes

Lysosomes are digestive bodies that break down foreign or damaged material in cells. A membrane surrounds each lysosome and separates its digestive enzymes from the rest of the cytoplasm.

Breaking it down

The lysosomal enzymes digest matter brought into the cell by *phagocytes*, special cells that surround and engulf matter outside the cell and then transport it through the cell membrane. The membrane of the lysosome fuses with the membrane of the cytoplasmic spaces surrounding the phagocytized material; this fusion allows the lysosomal enzymes to digest the engulfed material.

Plasma membrane

The *plasma membrane* (cell membrane) is the gatekeeper of the cell.
It serves as the cell's external boundary, separating it from other cells
and from the external environment.

Checkpoint

Nothing gets by this semipermeable membrane without authorization
from the nucleus. The semipermeable membrane consists of a double
layer of phospholipids with protein molecules.

Nucleus

The *nucleus* is the cell's mission control. It plays a role in cell growth,
metabolism, and reproduction.

A nucleus may contain one or more *nucleoli*—a dark-staining structure that synthesizes *ribonucleic acid* (RNA). The nucleus also contains
chromosomes. Chromosomes control cellular activity and direct protein
synthesis through ribosomes in the cytoplasm. (For more information
on chromosomes, see chapter 2, Genetics.)

DNA and RNA

Protein synthesis is essential for the growth of new tissue and the
repair of damaged tissue. *Deoxyribonucleic acid* (DNA) carries genetic
information and provides the blueprint for protein synthesis. RNA
transfers this genetic information to the ribosomes,
where protein synthesis occurs.

Touching all the bases

The basic structural unit of DNA is a *nucleotide*. Nucleotides consist of a phosphate group that's linked to a
five-carbon sugar, *deoxyribose*, and
joined to a nitrogen-containing
compound called a *base*. Four
different DNA bases exist:
1. adenine (A)
2. guanine (G)
3. thymine (T)
4. cytosine (C).

We're complementary. That means we fit together perfectly.

Identifying rings

Adenine and *guanine* are double-ring compounds classified as *purines. Thymine* and *cytosine* are single-ring compounds classified as *pyrimidines.*

The chain gangs

DNA chains exist in pairs held together by weak chemical attractions between the nitrogen bases on adjacent chains. Because of the chemical shape of the bases, adenine bonds only with thymine and guanine bonds only with cytosine. Bases that can link with each other are called *complementary.*

Insider trading

RNA consists of nucleotide chains that differ slightly from the nucleotide chains found in DNA. Several types of RNA are involved in the transfer (to the ribosomes) of genetic information essential to protein synthesis. (See *Types of RNA.*)

Now I get it!

Types of RNA

There are three types of ribonucleic acid (RNA): ribosomal, messenger, and transfer. Each has its own specific function.

Ribosomal RNA

Ribosomal RNA is used to make ribosomes in the endoplasmic reticulum of the cytoplasm, where the cell produces proteins.

Messenger RNA

Messenger RNA directs the arrangement of amino acids to make proteins at the ribosomes. Its single strand of nucleotides is complementary to a segment of the deoxyribonucleic acid chain that contains instructions for protein synthesis. Its chains pass from the nucleus into the cytoplasm, attaching to ribosomes there.

Transfer RNA

Transfer RNA consists of short nucleotide chains, each of which is specific for an individual amino acid. Transfer RNA transfers the genetic code from messenger RNA for the production of a specific amino acid.

Cell reproduction

Cells are under a constant call to reproduce; it's either that or die. Cell division is how cells reproduce (or replicate) themselves; they achieve this through the process of *mitosis* or *meiosis*.

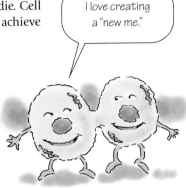

I love creating a "new me."

DNA does its thing

Before a cell divides, its chromosomes are duplicated. During this process, the double helix separates into two DNA chains. Each chain serves as a template for constructing a new chain. Individual DNA nucleotides are linked into new strands with bases complementary to those in the original.

Double double

In this way, two identical double helices are formed, each containing one of the original strands and a newly formed

Zoom in

DNA up close

Linked deoxyribonucleic acid (DNA) chains form a spiral structure, or *double helix.*

A spiral staircase
To understand linked DNA chains, imagine a spiral staircase. The deoxyribose and phosphate groups form the railings of the staircase, and the nitrogen base pairs (adenine and thymine, guanine and cytosine) form the steps.

Cell division
Each chain serves as a template for constructing a new chain. When a cell divides, individual DNA nucleotides are linked into new strands with bases complementary to those in the originals. In this way, two identical double helices are formed, each containing one of the original strands and a newly formed complementary strand. These double helices are duplicates of the original DNA chain.

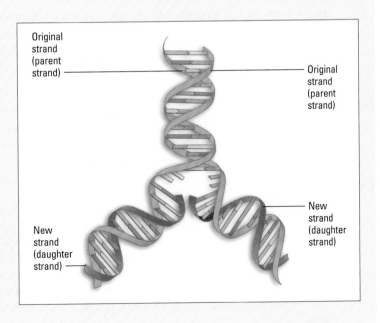

Original strand (parent strand)

Original strand (parent strand)

New strand (daughter strand)

New strand (daughter strand)

complementary strand. These double helices are duplicates of the original DNA chain. (See *DNA up close,* page 11.)

Mitosis

Mitosis is the equal division of material in the nucleus (*karyokinesis*) followed by division of the cell body (*cytokinesis*). It's the preferred mode of replication by all cells in the human body, except the gametes. Cell division occurs in five phases, an inactive phase called *interphase,* and four active phases:

1. prophase
2. metaphase
3. anaphase
4. telophase.

> He may think he's better than me, but we have identical DNA!

Two daughters equal 46

Mitosis results in two daughter cells (exact duplicates), each containing 23 pairs of chromosomes—or 46 individual chromosomes. This number is the *diploid number.* (See *Divide and conquer: five stages of mitosis.*)

Meiosis

Meiosis is reserved for gametes (ova and spermatozoa). This process intermixes genetic material between homologous chromosomes, producing four daughter cells, each with the *haploid number* of chromosomes (23, or half of the 46). Meiosis has two divisions separated by a resting phase.

First division

The first division has six phases and begins with one parent cell. When the first division ends, the result is two daughter cells—each containing the haploid (23) number of chromosomes.

Division 2, the sequel

The second division is a four-phase division that resembles mitosis. It starts with two new daughter cells, each containing the haploid number of chromosomes, and ends with four new haploid cells. In each cell, the two chromatids of each chromosome separate to form new daughter cells. However, because each cell entering the second division has only 23 chromosomes, each daughter cell formed has only 23 chromosomes. (See *Meiosis: Step-by-step,* page 14.)

Memory jogger

To help you remember the difference between haploid and diploid think of the prefix **di-** in diploid. Di- means **double**, so diploid cells have double the number of chromosomes in a haploid cell.

Now I get it!

Divide and conquer: five stages of mitosis

Through the process of mitosis, the nuclear content of all body cells (except gametes) reproduces and divides. The result is the formation of two new daughter cells, each containing the diploid (46) number of chromosomes.

Interphase

During *interphase*, the nucleus and nuclear membrane are well defined, and the nucleolus is visible. As chromosomes replicate, each forms a double strand that remains attached at the center by a centromere.

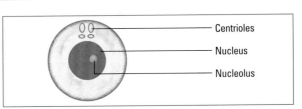

Centrioles

Nucleus

Nucleolus

Prophase

In *prophase*, the nucleolus disappears and the chromosomes become distinct. *Chromatids*, halves of each duplicated chromosome, remain attached by the centromere. Centrioles move to opposite sides of the cell and radiate spindle fibers.

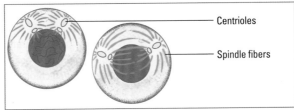

Centrioles

Spindle fibers

Metaphase

Metaphase occurs when chromosomes line up randomly in the center of the cell between the spindles, along the *metaphase plate*. The centromere of each chromosome then replicates.

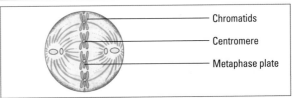

Chromatids

Centromere

Metaphase plate

Anaphase

Anaphase is characterized by centromeres moving apart, pulling the separate chromatids (now called *chromosomes*) to opposite ends of the cell. The number of chromosomes at each end of the cell equals the original number.

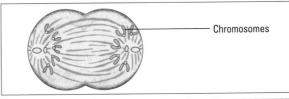

Chromosomes

Telophase

During *telophase*, the final stage of mitosis, a nuclear membrane forms around each nucleus and spindle fibers disappear. The cytoplasm compresses and divides the cell in half. Each new cell contains the diploid (46) number of chromosomes.

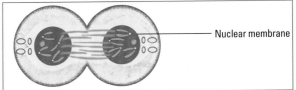

Nuclear membrane

Now I get it!

Meiosis: Step-by-step

Meiosis has two divisions that are separated by a resting phase. By the end of the first division, two daughter cells exist that each contains the haploid (23) number of chromosomes. When the second division ends, each of the two daughter cells from the first division divides, resulting in four daughter cells, each containing the haploid number of chromosomes.

First division

The first division has six phases. Here's what happens during each one.

Interphase

1. Chromosomes replicate, forming a double strand attached at the center by a centromere.
2. Chromosomes appear as an indistinguishable matrix within the nucleus.
3. Centrioles appear outside the nucleus.

Prophase I

1. The nucleolus and nuclear membrane disappear.
2. Chromosomes are distinct, with chromatids attached by the centromere.
3. Homologous chromosomes move close together and intertwine; exchange of genetic information (genetic recombination) may occur.
4. Centrioles separate and spindle fibers appear.

Metaphase I

1. Pairs of synaptic chromosomes line up randomly along the metaphase plate.
2. Spindle fibers attach to each chromosome pair.

Anaphase I

1. Synaptic pairs separate.
2. Spindle fibers pull homologous, double-stranded chromosomes to opposite ends of the cell.
3. Chromatids remain attached.

Telophase I

1. The nuclear membrane forms.
2. Spindle fibers and chromosomes disappear.
3. Cytoplasm compresses and divides the cell in half.
4. Each new cell contains the haploid (23) number of chromosomes.

Interkinesis

1. The nucleus and nuclear membrane are well defined.
2. The nucleolus is prominent, and each chromosome has two chromatids that don't replicate.

Second division

The second division closely resembles mitosis and is characterized by these four phases.

Prophase II

1. The nuclear membrane disappears.
2. Spindle fibers form.
3. Double-stranded chromosomes appear as thin threads.

Metaphase II

1. Chromosomes line up along the metaphase plate.
2. Centromeres replicate.

Anaphase II

1. Chromatids separate (now a single-stranded chromosome).
2. Chromosomes move away from each other to the opposite ends of the cell.

Telophase II

1. The nuclear membrane forms.
2. Chromosomes and spindle fibers disappear.
3. Cytoplasm compresses, dividing the cell in half.
4. Four daughter cells are created, each of which contains the haploid (23) number of chromosomes.

Cellular energy generation

All cellular function depends on energy generation and transportation of substances within and among cells.

Cellular power

Adenosine triphosphate (ATP) serves as the chemical fuel for cellular processes. ATP consists of a nitrogen-containing compound (adenine) joined to a five-carbon sugar (ribose), forming adenosine. Adenosine is joined to three phosphate (or triphosphate) groups. Chemical bonds between the first and second phosphate groups and between the second and third phosphate groups contain abundant energy.

The three R's

ATP needs to be converted to *adenosine diphosphate* (ADP) to produce energy. To understand this conversion, remember the three R's:

- *Rupture*—ATP is converted to ADP when the terminal high-energy phosphate bond ruptures.
- *Release*—Because the third phosphate is liberated, energy stored in the chemical bond is released.
- *Recycle*—Mitochondrial enzymes then reconvert ADP and the liberated phosphate to ATP. To obtain the energy needed for this reattachment, mitochondria oxidize food nutrients. This makes recycled ATP available again for energy production.

Movement within cells

Each cell interacts with body fluids through the interchange of substances.

Modes of transportation

Several transport methods—*diffusion, osmosis, active transport,* and *endocytosis*—move substances between cells and body fluids. In another method, *filtration,* fluids and dissolved substances are transferred across capillaries into *interstitial fluid* (fluid in the spaces between cells and tissues).

Diffusion

In *diffusion*, solutes move from an area of higher concentration to one of lower concentration. Eventually, an equal distribution of solutes between the two areas occurs.

Go with the flow

Diffusion is a form of passive transport—no energy is required to make it happen; it just happens. It's kind of like fish traveling downstream. They just go with the flow. (See *Understanding passive transport.*)

Advancing and declining rates

Several factors influence the rate of diffusion:
- *concentration gradient*—the greater the concentration gradient (the difference in particle concentration on either side of the plasma membrane), the faster the rate of diffusion

> No energy is required for diffusion so I'll just go with the flow.

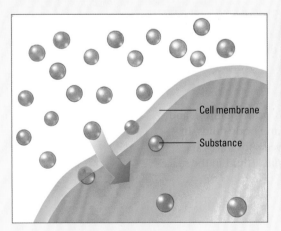

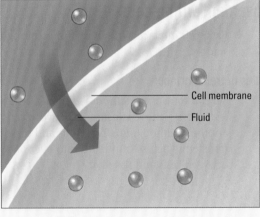

Now I get it!

Understanding passive transport

No energy is required for passive transport. It occurs through two mechanisms: diffusion and osmosis.

Diffusion
In diffusion, substances move from an area of higher concentration to an area of lower concentration. Movement continues until distribution is uniform.

Osmosis
In osmosis, fluid moves from an area of higher concentration to one of lower concentration.

Cell membrane

Substance

Cell membrane

Fluid

- *particle size*—the smaller the particles, the faster the rate of diffusion
- *lipid solubility*—the more lipid-soluble the particles are, the more rapidly they diffuse through the lipid layers of the cell membrane.

Several factors influence the rate of diffusion—concentration gradient, particle size, and lipid solubility.

Osmosis

Osmosis is the passive transport of fluid across a membrane, from an area of lower solute concentration (comparatively *more* fluid) into an area of higher solute concentration (comparatively *less* fluid).

Enough is enough

Osmosis stops when enough fluid has moved through the membrane to equalize the solute concentration on both sides of the membrane.

Active transport

Active transport requires energy. Usually, this mechanism moves a substance across the cell membrane against the concentration gradient—from an area of lower concentration to one of higher concentration. Think of active transport as swimming upstream. When a fish swims upstream, it has to expend energy.

ATP at it again

The energy required for a solute to move against a concentration gradient comes from ATP. ATP is stored in all cells and supplies energy for solute movement in and out of cells. (See *Understanding active transport*, page 18.)

Active transport requires energy. As a cell, that means I need to fill up on ATP.

It goes both ways

However, active transport also can move a substance with the concentration gradient. In this process, a carrier molecule in the cell membrane combines with the substance and transports it through the membrane, depositing it on the other side.

Endocytosis

Endocytosis is an active transport method in which, instead of passing through the cell membrane, a substance is engulfed by the cell. The cell surrounds the substance with part of the cell membrane. This part separates to form a *vacuole* (cavity) that moves to the cell's interior.

Now I get it!

Understanding active transport

Active transport moves molecules and ions against a concentration gradient from an area of lower concentration to one of higher concentration. This movement requires energy, usually in the form of adenosine triphosphate (ATP). The sodium-potassium pump and pinocytosis are examples of active transport mechanisms.

Sodium-potassium pump

The sodium-potassium pump moves sodium from inside the cell to outside, where the sodium concentration is greater; potassium moves from outside the cell to inside, where the potassium concentration is greater.

Pinocytosis

In pinocytosis, tiny vacuoles take droplets of fluid containing dissolved substances into the cell. The engulfed fluid is used in the cell.

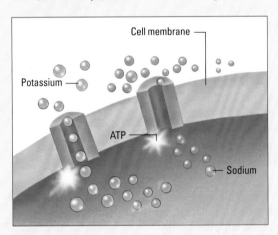

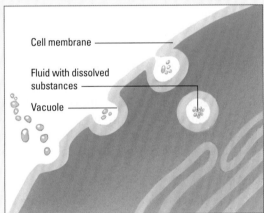

Gobbling up particles

Endocytosis involves either *phagocytosis* or *pinocytosis*. Phagocytosis refers to engulfment and ingestion of particles that are too large to pass through the cell membrane. Pinocytosis occurs only to engulf dissolved substances or small particles suspended in fluid.

Filtration

Fluid and dissolved substances also may move across a cell membrane by *filtration*.

Pressure is the point

In filtration, pressure (provided by capillary blood) is applied to a solution on one side of the cell membrane. The pressure forces fluid and dissolved particles through the membrane. The rate of filtration (how quickly substances pass through the membrane) depends on the amount of pressure. Filtration promotes the transfer of fluids and dissolved materials from the blood across the capillaries into the interstitial fluid.

A look at human tissue

Tissues are groups of cells that perform the same general function. The human body contains four basic types: *epithelial, connective, muscle,* and *nervous tissue.*

A crowd of us cells working together on the same function is called tissue.

Epithelial tissue

Epithelial tissue (epithelium) is a continuous cellular sheet that covers the body's surface, lines body cavities, and forms certain glands. Imagining a mummy wrapped in strips of cloth will give you some idea of how epithelial tissue covers the human body. (See *Distinguishing types of epithelial tissue,* page 20.)

Patrolling the borders

Some columnar epithelial cells in the lining of the intestines have vertical striations, forming a *striated border*. In the tubules of the kidneys, borders of columnar epithelial cells have tiny, brushlike structures (microvilli) called a *brush border*.

This hair isn't just for looks

Two common types of cells that form epithelial tissue are *stereociliated* and *ciliated epithelial cells*. The former line the epididymis and have long, piriform (pear-shaped) tufts. The latter possess *cilia,* fine hairlike protuberances. Cilia are larger than microvilli and move fluid and particles through the cavity of an organ.

Endothelium

Epithelial tissue with a single layer of squamous cells attached to a basement membrane is called *endothelium*. Such tissue lines the heart, lymphatic vessels, and blood vessels.

Distinguishing types of epithelial tissue

Epithelial tissue (epithelium) is classified by the number of cell layers and the shape of surface cells. Some types of epithelium go through a process of desquamation (shedding of debris) and regenerate continuously by transformation of cells from deeper layers.

Identified by number of cell layers

Classified by number of cell layers, epithelium may be *simple* (one-layered), *stratified* (multilayered), or *pseudostratified* (one-layered but appearing to be multilayered).

Classified by shape

If classified by shape, epithelium may be *squamous* (containing flat surface cells), *columnar* (containing tall, cylindrical surface cells), or *cuboidal* (containing cube-shaped surface cells).

The top left illustration below shows how the basement membrane of simple squamous epithelium joins the epithelium to underlying connective tissues. The remaining illustrations show the five other types of epithelial tissue.

Simple squamous epithelium
Single layer of flattened cells with disc-shaped nuclei

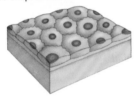

Simple columnar epithelium
Single layer of tall cells with oval nuclei

Stratified squamous epithelium
Basal cells that are cuboidal or columnar

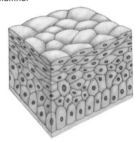

Simple cuboidal epithelium
Single layer of cubelike cells

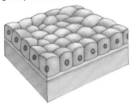

Stratified columnar epithelium
Superficial cells that are elongated and columnar

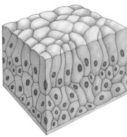

Pseudostratified columnar epithelium
Cells of different height with nuclei at different levels

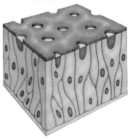

Glandular epithelium

Organs that produce secretions consist of a special type of epithelium called *glandular epithelium*.

The secret is in how it secretes

Glands are classified as endocrine or exocrine according to how they secrete their products.

- *Endocrine glands* release their secretions into the blood or lymph. For instance, the medulla of the adrenal gland secretes epinephrine and norepinephrine into the bloodstream.
- *Exocrine glands* discharge their secretions into ducts that lead to external or internal surfaces. For example, the sweat glands secrete sweat onto the surface of the skin.

Mixing it up

Mixed glands contain both endocrine and exocrine cells. The pancreas is a mixed gland. As an endocrine gland, it produces insulin and glucagon. As an exocrine gland, it introduces pancreatic juices into the intestines.

Connective tissue

Connective tissue—a category that includes bone, cartilage, and adipose (fatty) tissue—binds together and supports body structures. Connective tissue is classified as *loose* or *dense*.

Cut loose

Loose (areolar) connective tissue has large spaces that separate the fibers and cells. It contains a lot of intercellular fluid.

Dense tissue issues

Dense connective tissue provides structural support and has greater fiber concentration. Dense tissue is further subdivided into dense regular and dense irregular connective tissue:

- *Dense regular* connective tissue consists of tightly packed fibers arranged in a consistent pattern. It includes tendons, ligaments, and *aponeuroses* (flat fibrous sheets that attach muscles to bones or other tissues).
- *Dense irregular* connective tissue has tightly packed fibers arranged in an inconsistent pattern. It's found in the dermis, submucosa of the GI tract, fibrous capsules, and fasciae.

In addition to bones, connective tissue includes cartilage and adipose tissue.

Suppose it's adipose

Commonly called *fat, adipose tissue* is a specialized type of loose connective tissue where a single lipid (fat) droplet occupies most of each cell. Widely distributed subcutaneously, it acts as insulation to conserve body heat, as a cushion for internal organs, and as a storage depot for excess food and reserve supplies of energy. Adipose tissue also functions as endocrine tissue because it produces the appetite-suppressing hormone leptin.

What happened to your adipose tissue?

I went overboard on the dieting thing.

Muscle tissue

Muscle tissue consists of muscle cells with a generous blood supply. Muscle cells measure up to several centimeters long and have an elongated shape that enhances their *contractility* (ability to contract).

The tissues at issue

There are three basic types of muscle tissue:

1. *Striated muscle tissue* gets its name from its striped, or striated, appearance; it contracts voluntarily.
2. *Cardiac muscle tissue* is sometimes classified as striated because it's also composed of striated tissue. However, it differs from other striated muscle tissue in two ways: its fibers are separate cellular units that don't contain many nuclei, and it contracts involuntarily.
3. *Smooth muscle tissue* consists of long, spindle-shaped cells and lacks the striped pattern of striated tissue. Its activity is stimulated by the autonomic nervous system and isn't under voluntary control.

Wall tissue paper

Smooth muscle tissue lines the walls of many internal organs and other structures, including the respiratory passages from the trachea to the alveolar ducts, the urinary and genital ducts, the arteries and veins, the larger lymphatic trunks, the intestines, the arrectores pilorum, and the iris and ciliary body of the eye.

Nervous tissue

The main function of *nervous tissue* is communication. Its primary properties are *irritability* (the capacity to react to various physical and chemical agents) and *conductivity* (the ability to transmit the resulting reaction from one point to another).

Nervous tissue specialists

Neurons are highly specialized cells that generate and conduct nerve impulses. A typical neuron consists of a cell body with cytoplasmic extensions—numerous *dendrites* on one pole and a single *axon* on the opposite pole. These extensions allow the neuron to conduct impulses over long distances.

Irritable? I'll show you irritable!

Protecting neurons

Neuroglia form the support structure of nervous tissue, insulating and protecting neurons. They're found only in the central nervous system.

Quick quiz

1. The reference plane that divides the body lengthwise into right and left regions is the:
    A. frontal plane.
    B. sagittal plane.
    C. transverse plane.
    D. oblique plane.

Answer: B. Imaginary lines called *reference planes* are used to section the body. The sagittal plane runs lengthwise and divides the body into right and left regions.

2. The structure that plays the biggest role in cellular function is the:
    A. nucleus.
    B. Golgi apparatus.
    C. ribosome.
    D. mitochondrion.

Answer: A. Serving as the cell's control center, the nucleus plays a role in cell growth, metabolism, and reproduction.

3. The four basic types of tissue that the human body contains are:
    A. muscle, cartilage, glandular, and connective tissue.
    B. bone, cartilage, glands, and adipose tissue.
    C. loose, dense connective, dense regular, and dense irregular tissue.
    D. epithelial, connective, muscle, and nervous tissue.

Answer: D. Tissues are groups of cells with the same general function. The human body contains four basic types of tissue: epithelial, connective, muscle, and nervous tissue.

4. Meiosis ends when:
    A. two new daughter cells form, each with the haploid number of chromosomes.
    B. one daughter cell forms and is an exact copy of the original.
    C. four new daughter cells form, each with the haploid number of chromosomes.
    D. four new daughter cells form, each with the diploid number of chromosomes.

Answer: C. Meiosis comes to completion with the end of telophase II. The result is four daughter cells, each of which contains the haploid (23) number of chromosomes.

Scoring

☆☆☆   If you answered all four questions correctly, fantastic! You're well on your way to a fantastic voyage through the human body.

☆☆   If you answered three questions correctly, all right! You're in for some smooth sailing. Pretty soon you'll know the body inside and out.

☆   If you answered fewer than three questions correctly, buck up, camper. With plenty more quick quizzes to go, you'll conquer this body of knowledge in no time.

Selected References

Hall, J. (2015). *Guyton and Hall textbook of medical physiology* (13th ed.). Philadelphia, PA: Elsevier.

Saladin, K. (2014). *Anatomy & physiology: The unity of form and function* (7th ed.). New York, NY: McGraw Hill.

Chapter 2

Genetics

Just the facts

In this chapter, you'll learn:

◆ the way traits are transmitted
◆ the role of chromosomes and genes in heredity
◆ factors that determine trait predominance
◆ causes of genetic defects.

A look at genetics

Genetics is the study of heredity—the passing of traits from biological parents to their children. People inherit not only physical traits, such as eye color, but also biochemical and physiologic traits, including the tendency to develop certain diseases or conditions. Genetic information is carried in *genes,* which are strung together on the *deoxyribonucleic acid* (DNA) double helix to form chromosomes.

Family inheritance

Parents transmit inherited traits to their offspring in germ cells, or *gametes.* There are two types of human gametes: eggs (ova) and sperm (spermatozoa).

Chromosomes

The nucleus of each germ cell contains structures called *chromosomes.* Each chromosome contains a strand of genetic material called *DNA.* DNA is a long molecule that's made up of thousands of segments called *genes.* These genes carry the code for proteins that influence each trait a person inherits, ranging from blood type to toe shape. Chromosomes exist in pairs except in the germ cells.

Your genes determine how you look, how your body functions, and even whether you're prone to certain diseases.

Counting chromosomes

A human ovum contains 23 chromosomes. A sperm also contains 23 chromosomes, each similar in size and shape to a chromosome in the ovum. When an ovum and a sperm unite, the corresponding chromosomes pair up. The result is a fertilized cell with 46 chromosomes (23 pairs) in its nucleus.

Gen XX (or XY)

Of the 23 pairs of chromosomes in each living human cell, the two sex chromosomes of the 23rd pair determine a person's gender. The other 22 pairs are called *autosomes*.

In a female, both sex chromosomes are relatively large and each is designated by the letter X. In a male, one sex chromosome is an X chromosome and one is a smaller chromosome, designated by the letter Y.

Each gamete produced by a male contains either an X or a Y chromosome. Each gamete produced by a female contains an X chromosome. When a sperm with an X chromosome fertilizes an ovum, the offspring is female (two X chromosomes). When a sperm with a Y chromosome fertilizes an ovum, the offspring is male (one X and one Y chromosome).

Dividing the family assets

Ova and sperm are formed by a cell-division process called *meiosis*. In meiosis, each of the 23 pairs of chromosomes in a cell splits. The cell then divides, and each new cell (an ovum or sperm) receives one set of 23 chromosomes. (See chapter 1, The human body, page 14, for more information on meiosis.)

Genes

Genes are segments of a DNA chain, arranged in sequence on a chromosome. This sequence determines the properties of an organism.

Locus pocus

The location of a specific gene on a chromosome is called the *gene locus*. The locus of each gene is specific and doesn't vary from person to person. This allows each of the thousands of genes in an ovum to join the corresponding genes from a sperm when the chromosomes pair up at fertilization.

How do I look?

The genetic information stored at a locus of a gene determines the genetic constitution—or *genotype*—of a person. The detectable, outward manifestation of a genotype is called the *phenotype*.

The genome at a glance

The human genome is made up of a set of very long deoxyribonucleic acid (DNA) molecules, one for each chromosome. DNA has four different chemical building blocks, called *bases*. Each human genome contains about 3 billion of these bases, arranged in an order that's unique for each person. Increasing its complexity, arrayed along the DNA molecules are over 30,000 genes.

Consider its size: If the DNA sequence of the human genome were compiled in books, the equivalent of 200 volumes the size of the Manhattan telephone book (at 1,000 pages each) would be needed to hold it all.

If I were to read out loud the 3 billion bases in just one genome, it would take 9½ years.

The human genome

One complete set of chromosomes, containing all the genetic information for one person, is called a *genome*. (See *The genome at a glance*.)

The 411 on genes

For several years, scientists intensely studied the human genome to determine the entire sequence of each DNA molecule and the location and identity of all genes. The project was successfully completed in April 2003.

Catching the culprit

Genetic sequencing information allows practitioners to identify the causes of disease rather than simply treating symptoms. Other benefits include future development of more specific diagnostic tests, the formation of new therapies, and methods for avoiding conditions that trigger disease.

In addition, practitioners can now test patients for a gene error, present in 1 out of 500 people, that indicates an increased risk of developing colon cancer. Similarly, individuals with personal or family histories of breast or ovarian cancer can be tested for genetic predispositions to those diseases. Researchers are also seeking genes associated with dozens of other diseases, including chronic conditions such as asthma and diabetes.

It only takes one bad gene to spoil a gene pool party.

Trait predominance

Each parent contributes one set of chromosomes (and therefore one set of genes) to his or her offspring. Therefore, every offspring has two genes for every locus (location on the chromosome) on the autosomal chromosomes.

Variation is the spice of life

Some characteristics, or traits, are determined by one gene that may have many variants. Variations of the same gene are called *alleles*. A person who has identical alleles on each chromosome is *homozygous* for that trait; if the alleles are different, they're said to be *heterozygous*. Other traits—called *polygenic traits*—require the interaction of more than one gene.

Autosomal inheritance

On autosomal chromosomes, one allele may exert more influence in determining a specific trait. This is called the *dominant gene*. The less influential allele is called the *recessive gene*. Offspring express the trait of a dominant allele if both, or only one, chromosome in a pair carries it. For a recessive allele to be expressed, both chromosomes must carry recessive versions of the alleles. (See *How genes express themselves*.)

Sex-linked inheritance

The X and Y chromosomes are the sex chromosomes. The X chromosome is much larger than the Y. Therefore, males (who have XY chromosomes) have less genetic material than do females (who have

Now I get it!

How genes express themselves

Genes account for inherited traits. *Gene expression* refers to a gene's effect on cell structure or function; however, the effects vary with the gene.

Dominant genes

If genes could speak, dominant genes would be loud and garrulous, dominating every conversation! Dominant genes (such as the one for dark hair) can be expressed and transmitted to the offspring even if only one parent possesses the gene.

Recessive genes

Unlike dominant genes, recessive genes prefer to hide their light under a bushel basket. A recessive gene (such as the one for blond hair) is expressed only when both parents transmit it to the offspring.

Codominant genes

Firm believers in equality, codominant genes (such as the genes that direct specific types of hemoglobin synthesis in red blood cells) allow expression of both alleles.

Sex-linked genes

Sex-linked genes are carried on sex chromosomes. Almost all appear on the X chromosome and are recessive. In the male, sex-linked genes behave like dominant genes because no second X chromosome exists.

XX chromosomes), which means they have only one copy of most genes on the X chromosome. Inheritance of those genes is called *X-linked,* or *sex-linked, inheritance.*

Unequal X-change

A woman transmits one copy of each X-linked gene to each of her children, male or female. Because a man transmits an X chromosome only to his female children (male children receive a Y chromosome), he transmits X-linked genes only to his daughters, never his sons.

Multifactorial inheritance

Multifactorial inheritance reflects the interaction of at least two genes and the influence of environmental factors.

Raising the bar

Height is a classic example of a multifactorial trait. In general, the height of an offspring peaks between the heights of the two parents. However, nutritional patterns, health care, and other environmental factors also influence development of such traits as height. A better-nourished, healthier child of two short parents may be taller than either parent.

When it all comes together

Some diseases also have genetic predispositions for multifactorial inheritance; that is, the gene for a disease might be expressed only under certain environmental conditions.

Factors that may contribute to multifactorial inheritance include:
- use of drugs, alcohol, or hormones by either parent
- maternal smoking
- maternal or paternal exposure to radiation
- maternal infection during pregnancy
- preexisting diseases in the mother
- nutritional factors
- general maternal or paternal health
- maternal-fetal blood incompatibility
- inadequate prenatal care.

Good nutrition, proper health care, and other environmental factors can influence how a gene expresses itself.

Genetic defects

Genetic defects are defects that result from changes to genes or chromosomes. They're categorized as autosomal disorders, sex-linked disorders, or multifactorial disorders. Some defects arise spontaneously, whereas others may be caused by environmental teratogens. *Teratogens* are environmental agents (such as infectious toxins, maternal diseases, drugs, chemicals, and physical agents) that can cause structural or functional defects in a developing fetus. They may also cause spontaneous abortion, complications during labor and delivery, hidden defects in later development (such as cognitive or behavioral problems), and benign or cancerous tumors.

Permanent change in plans

A permanent change in genetic material is known as a *mutation*. Mutations can result from exposure to radiation, certain chemicals, or viruses. They may also happen spontaneously and can occur anywhere in the genome.

Every cell has built-in defenses against genetic damage. However, if a mutation isn't identified or repaired, it may produce a new trait that can be transmitted to offspring. Some mutations produce no effect, others change the expression of a trait, and others can even change the way a cell functions. Some mutations cause serious or deadly defects, such as congenital anomalies and cancer.

Autosomal disorders

In autosomal disorders, an error occurs at a single gene site on the DNA strand. Single-gene disorders are inherited in clearly identifiable patterns that are the same as those seen in inheritance of normal traits. Because every person has 22 pairs of autosomes and only 1 pair of sex chromosomes, most hereditary disorders are caused by autosomal defects.

The assertive type

Autosomal dominant transmission involves transmission of an abnormal gene that's dominant. Autosomal dominant disorders usually affect male and female offspring equally. Children with one affected parent have a 50% chance of being affected.

Passive-aggressive behavior

Autosomal recessive inheritance involves transmission of a recessive gene that's abnormal. Autosomal recessive disorders also usually affect male and female offspring equally. If both parents are affected, all their offspring will be affected. If both parents are unaffected but carry the defective gene, each child has a 25% chance of being affected. If only one parent is affected and the other isn't a carrier, none of the offspring will be affected, but all will carry the defective gene. If one parent is affected and the other is a carrier, 50% of their children will be affected. Because of this transmission pattern, autosomal recessive disorders may occur even when there's no family history of the disease.

Sex-linked disorders

Genetic disorders caused by genes located on the sex chromosomes are termed *sex-linked disorders*.

X calls the shots

Most sex-linked disorders are controlled by genes on the X chromosome, usually as recessive traits. Because males have only one X chromosome, a single X-linked recessive gene can cause disease to be exhibited in a male. Females receive two X chromosomes, so they may be homozygous for a disease allele (and exhibit the disease), homozygous for a normal allele (and neither have nor carry the disease), or heterozygous (carry, but not exhibit, the disease).

Most people who express X-linked recessive traits are males with unaffected parents. In rare cases, the father is affected and the mother is a carrier. All daughters of an affected male are carriers. An affected male never transmits the trait to his son. Unaffected male children of a female carrier don't transmit the disorder.

Evidence in history

With X-linked dominant inheritance, evidence of the inherited trait usually exists in the family history. A person with the abnormal trait must have one affected parent. If the father has an X-linked dominant disorder, all of his daughters and none of his sons will be affected. If a mother has an X-linked dominant disorder, each of her children has a 50% chance of being affected.

Most people who have X-linked recessive disorders are males. That's because males have only one X chromosome. Females have a second X chromosome that overpowers the "diseased" X.

Multifactorial disorders

Most multifactorial disorders result from a number of genes and environmental influences acting together. In polygenic inheritance, each gene has a small additive effect and the combination of genetic errors is unpredictable. Multifactorial disorders can result from a less-than-optimum expression of many different genes, not from a specific error.

Mixing it up

Some multifactorial disorders are apparent at birth, such as cleft lip, cleft palate, congenital heart disease, anencephaly, clubfoot,

Chromosomal disjunction and nondisjunction

This illustration shows disjunction and nondisjunction of an ovum. When disjunction proceeds normally, fertilization with a normal sperm results in a zygote with the correct number of chromosomes. In nondisjunction, the duplicating chromosomes fail to separate; the result is one trisomic cell and one monosomic cell.

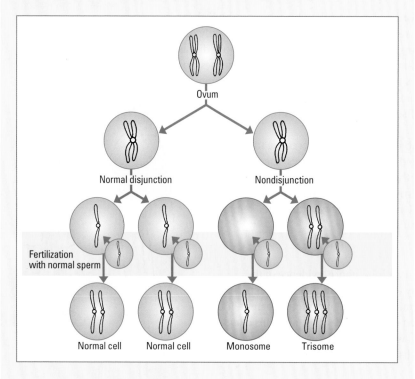

and myelomeningocele. Others, such as type II diabetes mellitus, hypertension, hyperlipidemia, most autoimmune diseases and many cancers, don't appear until later. Environmental factors most likely influence the development of multifactorial disorders during adulthood.

Chromosome defects

Aberrations in chromosome structure or number cause a class of disorders called *congenital anomalies,* or *birth defects.* Genetic aberrations include the loss, addition, or rearrangement of genetic material.

Most clinically significant chromosome aberrations arise during meiosis, an incredibly complex process that can go wrong in many ways. Potential contributing factors include maternal age, radiation, and use of some therapeutic or recreational drugs.

Unequally yoked

Translocation, the relocation of a segment of a chromosome to a non-homologous chromosome, occurs when chromosomes split apart and rejoin in an abnormal arrangement. The cells still have a normal amount of genetic material, so often there are no visible abnormalities. However, the children of parents with translocated chromosomes may have serious genetic defects, such as monosomies or trisomies.

A *monosomy* is a condition in which the number of chromosomes present is one less than normal; an autosomal monosomy is incompatible with life. The presence of an extra chromosome is called a *trisomy.* A mixture of both abnormal and normal cells results in *mosaicism* (two or more cell lines in the same person). The effects of mosaicism depend on the number and location of abnormal cells. Some common translocation outcomes are Down's syndrome and trisomy 13.

Breaking up is hard to do

During both meiosis and mitosis, chromosomes normally separate in a process called *disjunction.* Failure to separate, called *nondisjunction,* causes an unequal distribution of chromosomes between the two resulting cells. If nondisjunction occurs soon after fertilization, it may affect all the resulting cells. The incidence of nondisjunction increases with parental age. (See *Chromosomal disjunction and nondisjunction.*)

Quick quiz

1. What's the total number of chromosomes in a fertilized cell?
    A. 12
    B. 23
    C. 46
    D. 52

Answer: C. There are 46 chromosomes (23 pairs) in the nucleus of a fertilized cell.

2. According to genetic theory, if a child has cystic fibrosis (CF), this must mean:
    A. both parents transmit the gene for CF.
    B. one parent transmits the gene for CF.
    C. one grandparent has CF.
    D. neither parent has the gene for CF.

Answer: A. Because the trait for CF is a recessive gene, it's only expressed when both parents transmit it to the offspring.

3. The presence of an extra chromosome is called:
    A. monosomy.
    B. trisomy.
    C. mosaicism.
    D. nondisjunction.

Answer: B. Trisomy is the presence of an extra chromosome; it results from nondisjunction.

4. Which definition applies to the term *mutation?*
    A. An environmental agent responsible for a genetic defect
    B. A permanent change in genetic material
    C. Interaction of at least two abnormal genes
    D. Expression of a recessive gene in an offspring

Answer: B. A mutation is a permanent change in genetic material that may result from exposure to radiation, certain chemicals, or viruses. Mutations may also occur spontaneously.

5. A child has brown eyes and brown hair. This description reveals the child's:
    A. phenotype.
    B. genotype.
    C. genome.
    D. autosomes.

Answer: A. Phenotype refers to the outward, detectable manifestation of a person's genetic makeup or genotype.

Scoring

☆☆☆   If you answered all five questions correctly, excellent! When it comes to genetics, you're a gene-ius.

☆☆   If you answered four questions correctly, good job! Your understanding of genetics is definitely dominant.

☆   If you answered fewer than four questions correctly, don't worry. Another look at the chapter may help to X-plain Y.

Selected References

Hall, J. (2015). *Guyton and Hall textbook of medical physiology* (13th ed.). Philadelphia, PA: Elsevier.

Saladin, K. (2014). *Anatomy & physiology: The unity of form and function* (7th ed.). New York, NY: McGraw Hill.

Chapter 3

Chemical organization

Just the facts

In this chapter, you'll learn:

◆ the chemical composition of the body

◆ the structure of an atom

◆ differences between inorganic and organic compounds.

Without chemicals in the proper amounts, I would die.

A look at body chemistry

The human body is composed of chemicals; in fact, all of its activities are chemical in nature. To understand the human body and its functions, you must understand chemistry.

Comes down to chemistry

The chemical level is the simplest and most important level of structural organization. Without the proper chemicals in the proper amounts, body cells—and eventually the body itself—would die.

Principles of chemistry

Every cell contains thousands of different chemicals that constantly interact with one another. Differences in chemical composition differentiate types of body tissue. Furthermore, the blueprints of heredity (deoxyribonucleic acid [DNA] and ribonucleic acid [RNA]) are encoded in chemical form.

What's the matter?

Matter is anything that has mass and occupies space. It may be a solid, liquid, or gas.

Energetic types

Energy is the capacity to do work—to put mass into motion. It may be *potential energy* (stored energy) or *kinetic energy* (the energy of motion). Types of energy include chemical, electrical, and radiant.

Chemical composition

An *element* is matter that can't be broken down into simpler substances by normal chemical reactions. All forms of matter are composed of chemical elements. Each of the chemical elements in the periodic table has a chemical symbol. For example, N is the chemical symbol for nitrogen. (See *Understanding elements and compounds*.)

It's elementary

Carbon, hydrogen, nitrogen, and oxygen account for 96% of the body's total weight. Calcium and phosphorus account for another 2.5%. (See *What's a body made of?* page 38.)

Atomic structure

An *atom* is the smallest unit of matter that can take part in a chemical reaction. Atoms of a single type constitute an element.

Subatomic particles

Each atom has a dense central core called a *nucleus,* plus one or more surrounding energy layers called *electron shells.* Atoms consist of three basic subatomic particles: *protons, neutrons,* and *electrons.*

Weighing in

A proton weighs nearly the same as a neutron, and a proton and a neutron each weigh 1,836 times as much as an electron.

Protons

Protons (p+) are closely packed particles in the atom's nucleus that have a positive charge. Each element has a distinct number of protons.

Now I get it!

Understanding elements and compounds

It can be confusing to understand the difference between elements and compounds. The best way to remember the difference is to understand how atoms combine to form each.

Get at them atoms
A single atom constitutes an element. Thus, an atom of hydrogen is the element hydrogen. Now, here's the confusing part: An element can also be composed of more than one atom—a molecule.

Yep. I'm an atom—the smallest unit of matter that can take part in a chemical reaction.

Molecules: Two (or more) of the same
A molecule is a combination of two or more atoms. If these atoms are the same—that is, all the same element (such as all hydrogen atoms)—they're considered a *molecule* of that element (a molecule of hydrogen).

We're atoms that are joined together to make a molecule. We're the same, so we're a molecule of an element.

Compounds: Where the different come together
If the atoms combined are different—that is, different elements (such as a carbon atom and an oxygen atom)—the molecule formed is a *compound* (such as the compound CO, or carbon monoxide).

I know we just met, but I think we'll make a beautiful compound together.

Positive thinking

An element's number of protons determines its *atomic number* and positive charge. For example, all carbon atoms—and *only* carbon atoms—have six protons; therefore, the atomic number of carbon is 6 (6p+).

Body shop

What's a body made of?

This chart shows the chemical elements of the human body in descending order from most to least plentiful.

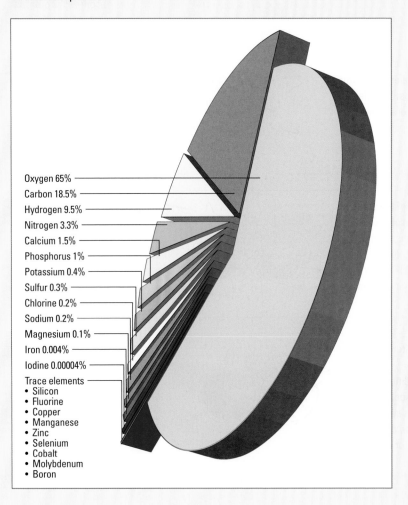

Oxygen 65%
Carbon 18.5%
Hydrogen 9.5%
Nitrogen 3.3%
Calcium 1.5%
Phosphorus 1%
Potassium 0.4%
Sulfur 0.3%
Chlorine 0.2%
Sodium 0.2%
Magnesium 0.1%
Iron 0.004%
Iodine 0.00004%
Trace elements
• Silicon
• Fluorine
• Copper
• Manganese
• Zinc
• Selenium
• Cobalt
• Molybdenum
• Boron

Neutrons

Neutrons (n) are uncharged, or neutral, particles in the atom's nucleus.

Mass numbers

An atom's *atomic mass number* is distinct from its atomic number. The atomic mass number is the sum of the number of protons and neutrons in the nucleus of an atom. You can also think of the atomic mass number as the sum of the masses of protons and neutrons. For example, helium, with two protons and two neutrons, has an atomic mass number of 4.

Isolating the isotopes

Not all the atoms of an element necessarily have the same number of neutrons. An *isotope* is a form of an atom that has a different number of neutrons and, therefore, a different atomic weight.

A weighty matter

Understanding isotopes is a key to another important concept, *atomic weight*. An atom's atomic weight is the average of the relative weights (atomic mass numbers) of all the element's isotopes. (Recall that isotopes are different atomic forms of the same element that vary in the number of neutrons they contain.)

Electrons

Electrons (e–) are negatively charged particles that orbit the nucleus in electron shells. They play a key role in chemical bonds and reactions.

Staying neutral

The number of electrons in an atom equals the number of protons in its nucleus. The electrons' negative charges cancel out the protons' positive charges, making atoms electrically neutral.

Shell games

Electrons circle the nucleus in *shells,* or concentric circles. Each electron shell can hold a maximum number of electrons and represents a specific energy level. The innermost shell can accommodate two electrons at most, whereas the outermost shells can hold many more.

An atom with single (unpaired) electrons orbiting in its outermost electron shell can be chemically *active*—that is, able to take part in chemical reactions. An atom with an outer shell that contains only pairs of electrons is chemically inactive, or *stable*.

The value of valence

An atom's valence (its ability to combine with other atoms) equals the number of unpaired electrons in its outer shell. For example, sodium (Na^+) has a plus-one valence because its outer shell contains an unpaired electron.

> It takes energy to bind atoms together. Breaking the bond releases energy.

Chemical bonds

A *chemical bond* is a force of attraction that binds a molecule's atoms together. Formation of a chemical bond usually requires energy. Breakup of a chemical bond usually releases energy.

The name's bond...

Several types of chemical bonds exist:
- A *hydrogen bond* occurs when two atoms associate with a hydrogen atom. Oxygen and nitrogen, for instance, commonly form hydrogen bonds.
- An *ionic* (electrovalent) *bond* occurs when valence electrons transfer from one atom to another.
- A *covalent bond* forms when atoms share pairs of valence electrons. (See *Picturing ionic and covalent bonds*.)

Chemical reactions

A *chemical reaction* involves unpaired electrons in the outer shells of atoms. In this reaction, one of two events occurs:
1. Unpaired electrons from the outer shell of one atom transfer to the outer shell of another atom.
2. One atom shares its unpaired electrons with another atom.

How will they react?

Energy, particle concentration, speed, and orientation determine whether a chemical reaction will occur. The four basic types of chemical reactions are *synthesis, decomposition, exchange,* and *reversible reactions.* (See *Comparing chemical reactions*, page 42.)

Now I get it!

Picturing ionic and covalent bonds

A chemical bond is a force of attraction that binds the atoms of a molecule together. Let's examine ionic and covalent bonds.

Ionic bonds

In an ionic bond, an electron is transferred from one atom to another. By forces of attraction, an electron is transferred from a sodium (Na) atom to a chlorine (Cl) atom. The result is a molecule of sodium chloride (NaCl).

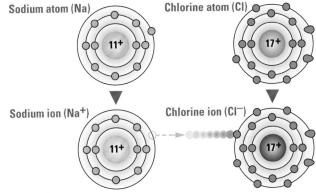

Sodium atom (Na)   Chlorine atom (Cl)

Sodium ion (Na$^+$)   Chlorine ion (Cl$^-$)

Covalent bonds

In a covalent bond, atoms share a pair of electrons. This is what happens when two hydrogen (H) atoms form a covalent bond.

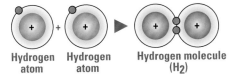

Hydrogen   Hydrogen   Hydrogen molecule
atom   atom   (H$_2$)

Comparing chemical reactions

When chemical reactions occur, they involve unpaired electrons in the outer shells of atoms. Here are the four basic types of chemical reactions.

Synthesis reaction (anabolism)

A synthesis reaction combines two or more substances (reactants) to form a new, more complex substance (product). This results in a chemical bond.

$$A + B \rightarrow A\,B$$

Decomposition reaction (catabolism)

In a decomposition reaction, a substance decomposes, or breaks down, into two or more simpler substances, leading to the breakdown of a chemical bond.

$$A\,B \rightarrow A + B$$

Exchange reaction

An exchange reaction is a combination of a decomposition and a synthesis reaction. This reaction occurs when two complex substances decompose into simpler substances. The simple substances then join (through synthesis) with different simple substances to form new complex substances.

$$A\,B + C\,D \rightarrow A + B + C + D \rightarrow A\,D + B\,C$$

Reversible reaction

In a reversible reaction, the product reverts to its original reactants, and vice versa. Reversible reactions may require special conditions, such as heat or light.

$$A + B \leftrightarrow A\,B$$

Inorganic and organic compounds

Although most biomolecules (molecules produced by living cells) form *organic compounds* (compounds containing carbon), some form *inorganic compounds* (compounds without carbon).

Inorganic compounds

Inorganic compounds are usually small and include water and *electrolytes*—inorganic acids, bases, and salts.

The body's reservoir

Water is the body's most abundant substance. It performs a host of vital functions, including:

- easily forming polar covalent bonds (which permits the transport of solvents)
- acting as a lubricant in mucus and other bodily fluids
- entering into chemical reactions, such as nutrient breakdown during digestion
- enabling the body to maintain a relatively constant temperature (by both absorbing and releasing heat slowly).

I need water to stay at a constant temperature.

Recognizing ionizing

Acids, bases, and salts are *electrolytes*—compounds whose molecules consist of positively charged ions, *cations,* and negatively charged ions, *anions,* that *ionize* (separate into ions) in solution:

- *Acids* ionize into hydrogen ions (H^+) and anions. In other words, acids separate into a positively charged hydrogen ion and a negatively charged anion.
- *Bases,* in contrast, ionize into hydroxide ions and cations. Bases separate into negatively charged hydroxide ions and positively charged cations.
- *Salts* form when acids react with bases. In water, salts ionize into cations and anions, but not hydrogen or hydroxide ions.

A balancing act

Body fluids must attain acid-base balance to maintain *homeostasis* (the dynamic equilibrium of the body). A solution's acidity is determined by the number of hydrogen ions it contains. The more hydrogen ions present, the more acidic the solution. Conversely, the more hydroxide ions a solution contains, the more basic, or *alkaline,* it is.

Organic compounds

Most biomolecules form *organic compounds*—compounds that contain carbon or carbon-hydrogen bonds. *Carbohydrates, lipids, proteins,* and *nucleic acids* are all examples of organic compounds.

I'm a carbohydrate. Count on me to release energy.

Carbohydrates

In the body, *carbohydrates* are sugars, starches, and glycogen.

The energy company

The main functions of carbohydrates are to release energy and store energy. There are three types of carbohydrates:

1. *Monosaccharides,* such as ribose and deoxyribose, are sugars with three to seven carbon atoms.
2. *Disaccharides,* such as lactose and maltose, contain two monosaccharides.
3. *Polysaccharides,* such as glycogen, are large carbohydrates with many monosaccharides.

Lipids

Lipids are water-insoluble biomolecules. The major lipids are *triglycerides, phospholipids, steroids, lipoproteins,* and *eicosanoids.*

To insulate and protect

Triglycerides are the most abundant lipid in both food and the body. These lipids are neutral fats that insulate and protect. They also serve as the body's most concentrated energy source. Triglycerides contain three molecules of a fatty acid chemically joined to one molecule of glycerol.

Bars on the cell

Phospholipids are the major structural components of cell membranes and consist of one molecule of glycerol, two molecules of a fatty acid, and a phosphate group.

No fat in cholesterol?

Steroids are simple lipids with no fatty acids in their molecules. They fall into four main categories, each of which performs different functions.
- *Bile salts* emulsify fats during digestion and aid absorption of the fat-soluble vitamins (vitamins A, D, E, and K).
- *Hormones* are chemical substances that have a specific effect on other cells.
- *Cholesterol,* a part of animal cell membranes, is needed to form all other steroids.
- *Vitamin D* helps regulate the body's calcium concentration.

Porters and other hardworking lipids

Lipoproteins help transport lipids to various parts of the body. *Eicosanoids* include *prostaglandins,* which, among other functions modify hormone responses, promote the inflammatory response, and open the airways, and *leukotrienes,* which also play a part in allergic and inflammatory responses.

Memory jogger

To recall the distinction between the three types of carbohydrates, remember the prefixes:

Mono- means one.

Di- means two; therefore, disaccharides contain two monosaccharides.

Poly- means many, so you can expect that polysaccharides contain many monosaccharides.

Bile salts aid absorption of us fat-soluble vitamins.

Proteins

Proteins are the most abundant organic compound in the body. They're composed of building blocks called *amino acids*. Amino acids are linked together by *peptide bonds*—chemical bonds that join the carboxyl group of one amino acid to the amino group of another.

I'm a protein. I'm built from blocks of amino acids that form polypeptides.

Building up the blocks

Many amino acids linked together form a *polypeptide*. One or more polypeptides form a protein. The sequence of amino acids in a protein's polypeptide chain dictates its shape. A protein's shape determines which of its many functions it performs:

- providing structure and protection
- promoting muscle contraction
- transporting various substances
- regulating processes
- serving as an enzyme (the largest group of proteins, which act as catalysts for crucial chemical reactions).

Nucleic acids

Nucleic acids' main function is to store and transmit genetic information. The nucleic acids DNA and RNA are composed of nitrogenous bases, sugars, and phosphate groups. The primary hereditary molecule, DNA, contains two long chains of deoxyribonucleotides, which coil into a double-helix shape.

Holding it together

Deoxyribose and phosphate units alternate in the "backbone" of the chains. Holding the two chains together are base pairs of adenine-thymine and guanine-cytosine.

RNA and its special function

Unlike DNA, RNA has a single-chain structure. It contains ribose instead of deoxyribose and replaces the base thymine with uracil. RNA transmits genetic information from the cell nucleus to the cytoplasm. In the cytoplasm, it guides protein synthesis from amino acids.

Quick quiz

1. The three most plentiful chemical elements in the human body are:
   A. phosphorus, hydrogen, and oxygen.
   B. carbon, oxygen, and silicon.
   C. oxygen, carbon, and hydrogen.
   D. oxygen, carbon, and nitrogen.

Answer: C. There are 22 chemical elements in the human body. Oxygen (65%), carbon (18.5%), and hydrogen (9.5%) are the three most plentiful.

2. Protons are closely packed particles in the atom's nucleus that have:
   A. a positive charge.
   B. a negative charge.
   C. a neutral charge.
   D. a mixed charge.

Answer: A. Protons are positively charged particles in the atom's nucleus.

3. An example of an organic compound is:
   A. water.
   B. an electrolyte.
   C. a protein.
   D. an acid.

Answer: C. Organic compounds are compounds that contain carbon. Examples include carbohydrates, lipids, proteins, and nucleic acids.

Scoring

☆☆☆ If you answered all three questions correctly, congratulations. You and this chapter have achieved homeostasis.

☆☆ If you answered two questions correctly, excellent. You and this chapter go together as neatly as amino acids forming a polypeptide chain.

☆ If you answered only one question correctly, no worries. Get yourself organized and then go back and review the chapter.

Selected References

Hall, J. (2015). *Guyton and Hall textbook of medical physiology* (13th ed.) Philadelphia, PA: Elsevier.

Saladin, K. (2014). *Anatomy & physiology: The unity of form and function* (7th ed.). New York, NY: McGraw Hill.

Chapter 4

Integumentary system

Just the facts

In this chapter, you'll learn:

◆ basic functions of the skin

◆ skin layers and their components

◆ the appendages (hair, nails, and glands) of the integumentary system.

A look at the integumentary system

The skin serves as the body's primary defense mechanism, protecting the body from invaders.

The integumentary system is the largest body system and includes the skin, or *integument*, and its appendages (the hair, nails, and certain glands).

Not just another pretty face

The integumentary system performs many vital functions, including:

- protection of inner body structures
- sensory perception
- regulation of body temperature
- excretion of some body fluids
- synthesis of vitamin D.

Protection

The skin maintains the integrity of the body surface by migration and shedding. It can repair surface wounds by intensifying normal cell replacement mechanisms. The skin's top layer, known as the *epidermis,* protects the body against noxious chemicals and invasion from pathogens.

Langerhans' cells to the rescue

Langerhans' cells are specialized cells within the epidermis. They enhance the body's immune response by helping lymphocytes process antigens entering the skin.

The skin's own sun block

Melanocytes, another type of skin cell, protect the skin by producing the brown pigment *melanin*, which helps filter ultraviolet (UV) light (irradiation). Exposure to UV light can stimulate melanin production.

Melanocytes protect the body from ultraviolet light.

Sensory perception

Sensory nerve fibers originate in the nerve roots along the spine and supply sensation to specific areas of the skin known as *dermatomes*.

Just sensational

These nerve fibers transmit various sensations, such as temperature, touch, pressure, pain, and itching, from the skin to the central nervous system. Autonomic nerve fibers carry impulses to smooth muscle in the walls of the skin's blood vessels, to the muscles around the hair roots, and to the sweat glands.

Body temperature regulation

Abundant nerves, blood vessels, and eccrine glands within the skin's deeper layer, the *dermis*, help to control body temperature (thermoregulation).

Warming up...

When the skin is exposed to cold or internal body temperature falls, blood vessels constrict, decreasing blood flow and thereby conserving body heat.

...and cooling down

If the skin becomes too hot or internal body temperature rises, small arteries within the skin dilate, increasing blood flow, which in turn reduces body heat. (See *The skin's role in thermoregulation*.)

Excretion

The skin is also an excretory organ. The sweat glands excrete sweat, which contains water, electrolytes, urea, and lactic acid.

Now I get it!

The skin's role in thermoregulation

Abundant nerves, blood vessels, and eccrine glands within the skin's deeper layer aid thermoregulation (control of body temperature). The first part of the flow chart shows how the body conserves body heat. The second part of the flow chart shows how the body reduces body heat. Here's how the skin does its job.

It's time to warm up
The skin becomes exposed to cold or internal body temperature falls.

Blood vessels constrict in response to stimuli from the autonomic nervous system.

Blood flow decreases through the skin and body heat is conserved.

Now let's cool things off
The skin becomes too hot or internal body temperature rises.

Increased blood flow reduces body heat. If this doesn't lower temperature, the eccrine glands act to increase sweat production, and evaporation cools the skin.

Small arteries in the second skin layer dilate (expand).

Water works

While it eliminates body wastes through its more than two million pores, the skin also prevents body fluids from escaping. Here, the skin protects the body by preventing dehydration caused by loss of internal body fluids—as well as maintaining these levels by regulating the content and volume of sweat. It also keeps unwanted fluids in the environment from entering the body.

Vitamin D synthesis

Ultraviolet light from the sun activates a precursor compound located in the cells of the skin, producing vitamin D.

Skin layers

Two distinct layers of skin, the *epidermis* and *dermis,* lie above a third layer of *subcutaneous tissue*—sometimes called the *hypodermis.* (See *A close look at skin.*)

Zoom in

A close look at skin

Major components of skin include the epidermis, dermis, and epidermal appendages.

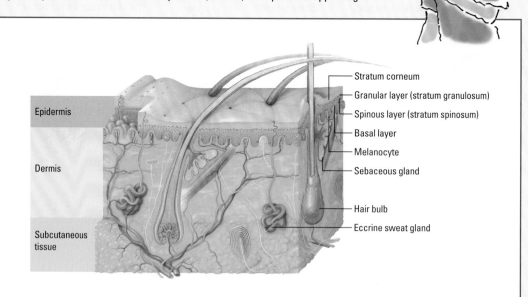

Epidermis

Dermis

Subcutaneous tissue

Stratum corneum

Granular layer (stratum granulosum)

Spinous layer (stratum spinosum)

Basal layer

Melanocyte

Sebaceous gland

Hair bulb

Eccrine sweat gland

Epidermis

The *epidermis* is the outermost layer and varies in thickness from less than 0.1 mm on the eyelids to more than 1 mm on the palms and soles. It's translucent, meaning it allows light to pass partially through it.

The ins and outs (and in betweens)

The epidermis is composed of avascular, stratified, squamous (scaly or platelike) epithelial tissue and is divided into five distinct layers. Each layer is named for its structure or function:

- The *stratum corneum,* or *horny layer,* is the outermost layer and consists of tightly arranged layers of cellular membranes and keratin.
- The *stratum lucidum,* or *clear layer,* blocks water penetration or loss. It may be missing in some thin skin.
- The *stratum granulosum,* or *granular layer,* is responsible for keratin formation and, like the stratum lucidum, may be missing in some thin skin.
- The *stratum spinosum,* or *spiny layer,* also helps with keratin formation and is rich in ribonucleic acid.
- The *stratum basale,* or the *basal layer,* is the innermost layer and produces new cells to replace the superficial keratinized cells that are continuously shed or worn away.

No blood, just rete pegs

The epidermis doesn't contain blood vessels. Food, vitamins, and oxygen are transported to this layer through fingerlike structures called *rete pegs,* which contain a network of tiny blood vessels. Rete pegs project down from the epidermis and up through the dermis, increasing contact between the layers.

Dermis

The *dermis,* also called the *corium,* is the skin's second layer. It's an elastic system that contains and supports blood vessels, lymphatic vessels, nerves, and the epidermal appendages.

What's in the matrix?

Most of the dermis is made up of extracellular material called *matrix.* Matrix contains:

- *collagen,* a protein made by the fibroblasts that gives strength and resilience to the dermis
- *elastic fibers* that bind the collagen and make the skin flexible.

Memory jogger

You can keep straight which skin layer is which by remembering that the prefix **epi**- means "upon." Therefore, the **epi**dermis is upon, or on top of, the dermis.

The layered look

The dermis itself has two layers:

- The *papillary dermis* has fingerlike projections, *papillae,* that connect the dermis to the epidermis. It contains characteristic ridges that on the fingers are known as *fingerprints.* These ridges also help the fingers and toes in gripping surfaces.
- The *reticular dermis* covers a layer of subcutaneous tissue. It's made of collagen fibers and provides strength, structure, and elasticity to the skin.

Subcutaneous tissue

Beneath the dermis is the third layer—*subcutaneous tissue*—which is a layer of fat. It contains larger blood vessels and nerves, as well as adipose cells, which are filled with fat. This subcutaneous fat layer lies on the muscles and bones. Functions of subcutaneous tissue include insulation, shock absorption, and storage of energy reserves.

Epidermal appendages

Numerous epidermal appendages occur throughout the skin. They include the hair, nails, sebaceous glands, and sweat glands. (See *Skin, hair, and nail changes with aging.*)

Hair

Hairs are long, slender shafts composed of keratin. At the expanded lower end of each hair is a bulb or root. On its undersurface, the root is indented by a *hair papilla,* a cluster of connective tissue and blood vessels.

It will literally make your hair stand on end

Each hair lies within an epithelium-lined sheath called a *hair follicle.* A bundle of smooth muscle fibers, *arrector pili,* extends through the dermis to attach to the base of the follicle. When these muscles contract, hair stands on end. Hair follicles also have a rich blood and nerve supply.

When arrector pili muscles contract, hair stands on end.

Skin, hair, and nail changes with aging

In the integumentary system, age-related changes can involve the skin, hair, and nails.

The skinny

As people age, their skin changes. For example, they may notice lines around their eyes (crow's feet), mouth, and nose. These lines result from subcutaneous fat loss, dermal thinning, decreasing collagen and elastin, and a 50% decline in cell replacement. Women's skin shows signs of aging about 10 years earlier than does men's because it's thinner and drier.

Because of the decreased rate of skin cell replacement, wounds may heal more slowly and be prone to infection in older people. In very old people, skin loses its elasticity and may seem almost transparent.

Other changes include drying of mucous membranes, which is caused by decreased sweat gland output and number of active sweat glands. This decrease in size, number, and function of sweat glands, combined with a loss of subcutaneous fat, makes it more difficult to regulate body temperature.

Melanocyte production also decreases as a person ages; however, melanocytes often proliferate in localized areas, causing brown spots (senile lentigo). This typically occurs in areas regularly exposed to the sun. Other common skin conditions in older people include senile keratosis (dry, harsh skin) and senile angioma (a benign tumor of dilated blood vessels caused by weakened capillary walls).

Hairy situation

Hair changes also occur with aging. Hair pigment decreases and hair may turn grey or white. This loss of pigment makes hair thinner; by age 70, it's baby fine again. Hormonal changes cause pubic hair loss. At the same time, facial hair commonly increases in postmenopausal women and decreases in aging men.

Nailed down

Aging may also alter nails. They may grow at different rates, and longitudinal ridges, flaking, brittleness, and malformations may increase. Toenails may also discolor and become thicker.

Nails

The *nails* are situated over the distal surface of the end of each finger and toe. Nails are composed of a specialized type of keratin.

On a bed of nails

The *nail plate,* surrounded on three sides by the nail folds, or *cuticles,* lies on the nail bed. The nail plate is formed by the nail matrix, which extends proximally for about ¼" (0.5 cm) beneath the nail fold.

Landing on the lunula

The distal portion of the matrix shows through the nail as a pale crescent-moon–shaped area. This is called the *lunula.* The translucent nail plate distal to the lunula exposes the nail bed. The vascular bed imparts the characteristic pink appearance under the nails. (See *A nail on a large scale,* page 54.)

Zoom in

A nail on a large scale

This illustration shows the anatomic components of a fingernail.

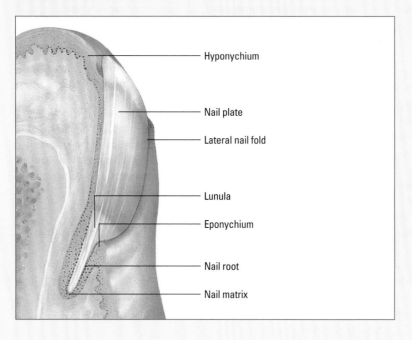

- Hyponychium
- Nail plate
- Lateral nail fold
- Lunula
- Eponychium
- Nail root
- Nail matrix

Sebaceous glands

Sebaceous glands are part of the hair follicle and occur on all parts of the skin except the palms and soles. They're most prominent on the scalp, face, upper torso, and genitalia.

A (small) miracle oil

The sebaceous glands produce *sebum,* a mixture of keratin, fat, and cellulose debris. Combined with sweat, sebum forms a moist, oily, acidic film that's mildly antibacterial and antifungal and that protects the skin surface. Sebum exits through the hair follicle opening to reach the skin surface.

Sweat glands

There are two types of sweat glands: eccrine glands and apocrine glands.

Eccrine glands

The *eccrine glands* are widely distributed throughout the body and produce an odorless, watery fluid with a sodium concentration equal to that of plasma. A duct from the coiled secretory portion passes through the dermis and epidermis, opening onto the skin surface.

Stressed out

Eccrine glands in the palms and soles secrete fluid mainly in response to emotional stress. For example, your eccrine glands might secrete fluid while you're taking a test. The remaining three million eccrine glands respond primarily to thermal stress, effectively regulating temperature. Eccrine glands are found everywhere except the lips and glans penis.

Apocrine glands

The *apocrine glands* are located chiefly in the axillary (underarm) and anogenital (groin) areas. They have a coiled secretory portion that lies deeper in the dermis than does that of the eccrine glands. A duct connects an apocrine gland to the upper portion of the hair follicle.

Oh no, b.o.

Apocrine glands begin to function at puberty. However, they have no known biological function. As bacteria decompose the fluids produced by these glands, body odor occurs.

Eccrine glands secrete fluid in response to stress.

Quick quiz

1. The main functions of the skin include:
    A. support, nourishment, and sensation.
    B. protection, sensory perception, and temperature regulation.
    C. fluid transport, sensory perception, and aging regulation.
    D. protection, motor response, and filtration.

Answer: B. The skin's main functions involve protection from injury, noxious chemicals, and bacterial invasion; sensory perception of touch, temperature, and pain; and regulation of body heat.

2. The outermost layer of the skin is the:
    A. epidermis.
    B. dermis.
    C. hypodermis.
    D. papillary dermis.

Answer: A. The outermost layer of the skin, composed of avascular, stratified, and squamous epithelial tissue, is the epidermis.

3. Which integumentary system structure is considered an epidermal appendage?
   A. Blood vessel
   B. Nerve
   C. Stratum basale
   D. Hair

Answer: D. The appendages of the epidermis are the nails, hair, sebaceous glands, eccrine glands, and apocrine glands.

4. Sebum is a mixture of:
   A. cellulose debris, fat, and keratin.
   B. collagen and elastin.
   C. watery fluid and sodium.
   D. protein, water, and electrolytes.

Answer: A. Sebum is produced by the sebaceous glands and is a mixture of keratin, fat, and cellulose debris.

5. The sweat glands that are widely distributed throughout the body are:
   A. apocrine.
   B. eccrine.
   C. adipose.
   D. sebaceous.

Answer: B. Eccrine glands are widely distributed throughout the body and produce an odorless, watery fluid.

Scoring

☆☆☆   If you answered all five questions correctly, amazing! You've got the integumentary system covered.

 ☆☆   If you answered four questions correctly, awesome. You're scratching beneath the surface of this body system.

  ☆   If you answered fewer than four questions correctly, no sweat. Just go back and review the chapter.

Selected References

Hall, J. (2015). *Guyton and Hall textbook of medical physiology* (13th ed.) Philadelphia, PA: Elsevier.

Saladin, K. (2014). *Anatomy & physiology: The unity of form and function* (7th ed.). New York, NY: McGraw Hill.

Chapter 5

Musculoskeletal system

Just the facts

In this chapter, you'll learn:

◆ major muscles and bones of the body

◆ types of muscle tissue and their functions

◆ types of bones and their functions

◆ the roles of tendons, ligaments, cartilage, joints, and bursae in body movement and structure.

A look at the musculoskeletal system

The musculoskeletal system consists of muscles, tendons, ligaments, bones, cartilage, joints, and bursae. These structures give the human body its shape and ability to move.

How the body moves

Various parts of the musculoskeletal system work with the nervous system to produce voluntary movements. Muscles contract when stimulated by impulses from the nervous system.

Using the force

During contraction, the muscle shortens, pulling on the bones to which it's attached. Force is applied to the tendon; then one bone is pulled toward, moved away from, or rotated around a second bone, depending on the type of muscle that has contracted. Most movement involves groups of muscles rather than one muscle.

> Structures of the musculoskeletal system work together to provide support and produce movement.

Muscles

There are three major types of muscle in the human body. They're classified by the tissue they contain:

1. *cardiac* (heart) muscle, which is made up of a specialized type of striated tissue
2. *smooth* (involuntary) muscle, which contains smooth muscle tissue
3. *skeletal* (voluntary and reflex) muscle, which consists of striated tissue.

The attached type

This chapter discusses only skeletal muscle—the type attached to bone. The human body has about 600 skeletal muscles. (See *Viewing the major skeletal muscles*.)

Memory jogger

To remember the functions of muscles, think miles per hour, or **MPH**:

Movement

Posture

Heat.

Muscle functions

Skeletal muscles move body parts or the body as a whole. They're responsible for both voluntary and reflex movements. Skeletal muscles also maintain posture and generate body heat.

Muscle structure

Skeletal muscle is composed of large, long cells called *muscle fibers*. Each fiber has many nuclei and a series of increasingly smaller internal fibrous structures. (See *Muscle structure up close*, page 60.)

Outside in

The structures of a muscle fiber, working from the cell's exterior to its interior, are:

- *endomysium*—the connective tissue layer surrounding an individual skeletal muscle fiber
- *sarcolemma*—the plasma membrane of the cell that lies beneath the endomysium and just above the cell's nucleus
- *sarcoplasm*—the muscle cell's cytoplasm, which is contained within the sarcolemma
- *myofibrils*—tiny, threadlike structures that run the fiber's length and make up the bulk of the fiber
- *myosin* (thick filaments) and *actin* (thin filaments)—still finer fibers within the myofibrils; there are about 1,500 myosin and about 3,000 actin.

Viewing the major skeletal muscles

This illustration shows anterior and posterior views of some of the major muscles.

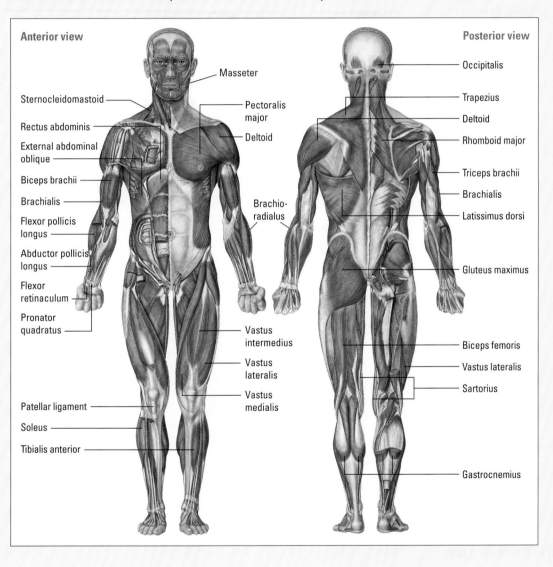

Anterior view

- Masseter
- Sternocleidomastoid
- Pectoralis major
- Rectus abdominis
- Deltoid
- External abdominal oblique
- Biceps brachii
- Brachialis
- Brachio-radialus
- Flexor pollicis longus
- Abductor pollicis longus
- Flexor retinaculum
- Pronator quadratus
- Vastus intermedius
- Vastus lateralis
- Vastus medialis
- Patellar ligament
- Soleus
- Tibialis anterior

Posterior view

- Occipitalis
- Trapezius
- Deltoid
- Rhomboid major
- Triceps brachii
- Brachialis
- Latissimus dorsi
- Gluteus maximus
- Biceps femoris
- Vastus lateralis
- Sartorius
- Gastrocnemius

Muscle structure up close

Skeletal muscle contains cell groups called *muscle fibers*. This illustration shows the muscle and its fibers.

In a bind

The *perimysium*—a sheath of connective tissue—binds muscle fibers together into a bundle (fascicle). The *epimysium* binds the fascicles together; beyond the muscle, it becomes a tendon.

Surrounded

A sarcolemma is a thin membrane enclosing a muscle fiber. Tiny myofibrils within the muscle fibers contain even finer fibers called *myosin* (thick filaments) and *actin* (thin filaments).

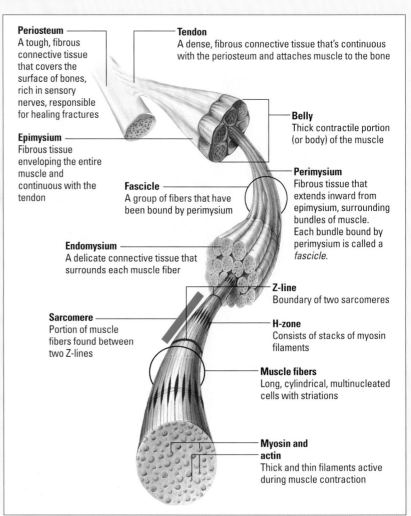

Periosteum
A tough, fibrous connective tissue that covers the surface of bones, rich in sensory nerves, responsible for healing fractures

Epimysium
Fibrous tissue enveloping the entire muscle and continuous with the tendon

Fascicle
A group of fibers that have been bound by perimysium

Endomysium
A delicate connective tissue that surrounds each muscle fiber

Sarcomere
Portion of muscle fibers found between two Z-lines

Tendon
A dense, fibrous connective tissue that's continuous with the periosteum and attaches muscle to the bone

Belly
Thick contractile portion (or body) of the muscle

Perimysium
Fibrous tissue that extends inward from epimysium, surrounding bundles of muscle. Each bundle bound by perimysium is called a *fascicle*.

Z-line
Boundary of two sarcomeres

H-zone
Consists of stacks of myosin filaments

Muscle fibers
Long, cylindrical, multinucleated cells with striations

Myosin and actin
Thick and thin filaments active during muscle contraction

Sarcomeres end to end

Myosin and actin are contained within compartments called *sarcomeres*. Sarcomeres are the functional units of skeletal muscle. During muscle contraction, myosin and actin slide over each other, reducing sarcomere length.

Zebra stripes

The sarcomere compartments of all the myofibrils in a single fiber are aligned. When a muscle fiber is viewed microscopically, transverse (at right angles to the long axis) stripes, called *striations* (or Z-lines), appear along the length of the fiber. Z-lines mark the beginning of sarcomeres.

A bundle of bundles

A fibrous sheath of connective tissue, called the *perimysium*, binds muscle fibers into a bundle, or *fascicle*. A stronger sheath, the *epimysium*, binds all of the fascicles together to form the entire muscle. Extending beyond the muscle, the epimysium becomes a tendon.

When I contract my muscles to draw my bow, one bone stays stationary. The other bone is pulled toward the stationary one.

Muscle attachment

Most skeletal muscles are attached to bones, either directly or indirectly.

The direct approach

In a direct attachment, the epimysium of the muscle fuses to the *periosteum*, the fibrous membrane covering the bone.

Being indirect

In an indirect attachment (most common), the epimysium extends past the muscle as a tendon, or *aponeurosis*, and attaches to the bone.

Contraction

During contraction, one of the bones to which the muscle is attached stays relatively stationary while the other is pulled in toward the stationary one.

Origin and insertion

The point where the muscle attaches to the stationary or less movable bone is called the *origin*; the point where it attaches to the more movable bone is called the *insertion*. The origin usually lies on the proximal end of the bone. The insertion site is on the distal end.

Factors such as genetic constitution and exercise cause muscle strength and size to differ among individuals.

Muscle growth

Muscle develops when existing muscle fibers hypertrophy. Muscle strength and size differ among individuals because of such factors as exercise, nutrition, gender, age, and genetic constitution. Changes in nutrition or exercise affect muscle strength and size in an individual. (See *Musculoskeletal changes with aging.*)

Muscle movements

Skeletal muscle can permit several types of movement. A muscle's functional name comes from the type of movement it permits. For example, a flexor muscle permits bending *(flexion)*; an adductor muscle permits movement toward a body axis *(adduction)*; and a circumductor muscle allows a circular movement *(circumduction).* (See *Basics of body movement.*)

Muscles of the axial skeleton

The muscles of the axial skeleton are essential for respiration, speech, facial expression, posture, and chewing. They include:
- muscles of the face, tongue, and neck
- muscles of mastication (chewing)
- muscles of the vertebral column situated along the spine
- muscles of the rib cage.

 Senior moment

Musculoskeletal changes with aging

As an individual ages, an apparent musculoskeletal change is a decrease in height. This occurs because exaggerated spinal curvature and narrowed intervertebral spaces cause the trunk to shorten and the arms to appear relatively long.

Other musculoskeletal changes that occur with aging include decreased muscle mass, which may result in muscle weakness, and decreased bone density, which causes bones to fracture more readily. In addition, collagen formation declines, which causes joints and supporting structures to lose resilience and elasticity. Synovial fluid also becomes more viscous, and synovial membranes become more fibrotic, making joints stiff.

Aging may also make tandem walking difficult. Usually, elderly people walk with shorter steps and wider leg stances for better balance and stability.

 **Body shop**

Basics of body movement

Basic muscle movement is best demonstrated in the diarthrodial joints, which allow 13 angular and circular movements:

- The shoulder demonstrates circumduction.
- The elbow demonstrates flexion and extension.
- The hip demonstrates internal and external rotation.
- The arm demonstrates abduction and adduction.
- The hand demonstrates supination and pronation.
- The jaw demonstrates retraction and protraction.
- The foot demonstrates eversion and inversion.

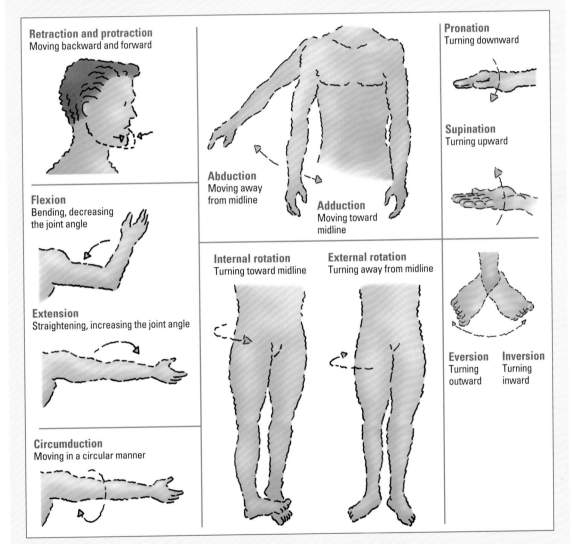

Retraction and protraction
Moving backward and forward

Flexion
Bending, decreasing the joint angle

Extension
Straightening, increasing the joint angle

Circumduction
Moving in a circular manner

Abduction
Moving away from midline

Adduction
Moving toward midline

Internal rotation
Turning toward midline

External rotation
Turning away from midline

Pronation
Turning downward

Supination
Turning upward

Eversion
Turning outward

Inversion
Turning inward

Muscles of the appendicular skeleton

The appendicular skeleton includes the muscles of the:
- shoulder
- abdominopelvic cavity
- upper and lower extremities.

Muscles of the upper extremities are classified according to the bones they move. Those that move the arm are further categorized into those with an origin on the axial skeleton and those with an origin on the scapula.

Tendons and ligaments

Tendons are bands of fibrous connective tissue that attach muscles to the periosteum, the fibrous covering of the bone. Tendons enable bones to move when skeletal muscles contract.

Ligaments are dense, strong, flexible bands of fibrous connective tissue that bind bones to other bones.

Bones

The human skeleton contains 206 bones: 80 form the *axial skeleton*—called *axial* because it lies along the central line, or axis, of the body—and 126 form the *appendicular skeleton*—relating to the limbs, or appendages, of the body. (See *Viewing the major bones*.)

Access the axis

Bones of the axial skeleton include:
- facial and cranial bones
- hyoid bone
- vertebrae
- ribs and sternum.

Appendages to the axis

Bones of the appendicular skeleton include:
- clavicle
- scapula
- humerus, radius, ulna, carpals, metacarpals, and phalanges
- pelvic bones
- femur, patella, fibula, tibia, tarsals, metatarsals, and phalanges.

Body shop

Viewing the major bones

These illustrations show the major bones and bone groups in the body.

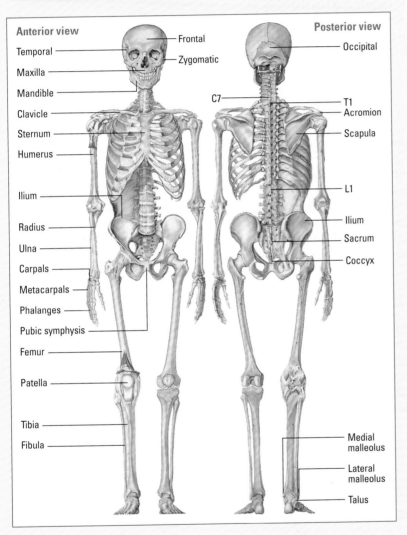

Anterior view

- Temporal
- Maxilla
- Mandible
- Clavicle
- Sternum
- Humerus
- Ilium
- Radius
- Ulna
- Carpals
- Metacarpals
- Phalanges
- Pubic symphysis
- Femur
- Patella
- Tibia
- Fibula

Frontal
Zygomatic

C7

Posterior view

Occipital

T1
Acromion
Scapula

L1

Ilium
Sacrum
Coccyx

Medial malleolus
Lateral malleolus
Talus

Bone classification

Bones are typically classified by shape. Thus, bones may be classified as:

- long (such as the humerus, radius, femur, and tibia) (See *Viewing a long bone.*)
- short (such as the carpals and tarsals)
- flat (such as the scapula, ribs, and skull)
- irregular (such as the vertebrae and mandible)
- sesamoid, which is a small bone developed in a tendon (such as the patella).

Bone functions

Bones perform various anatomic (mechanical) and physiologic functions. They:

- protect internal tissues and organs
- stabilize and support the body
- provide a surface for muscle, ligament, and tendon attachment
- move through "lever" action when contracted
- produce red blood cells in the bone marrow (*hematopoiesis*)
- store mineral salts (such as 99% of the body's calcium).

I support and stabilize the body.

Blood supply

Blood reaches bones through three paths:

1. *haversian canals*, minute channels that lie parallel to the axis of the bone and are passages for arterioles
2. *Volkmann's canals*, which contain vessels that connect one haversian canal to another and to the outer bone
3. *vessels* in the bone ends and within the marrow.

Bone formation

At 3 months in utero, the fetal skeleton is composed of cartilage. By about 6 months, fetal cartilage has been transformed into bony skeleton. (See *Bone growth and remodeling*, pages 68 and 69.)

Ossification is hard work

After birth, some bones—most notably the carpals and tarsals—*ossify* (harden). The change results from *endochondral ossification*, a process by which *osteoblasts* (bone-forming cells) produce *osteoid* (a collagenous material that ossifies).

(Text continues on page 70)

Zoom in

Viewing a long bone

The main parts of a long bone, shown below left, are the *diaphysis* (shaft) and the *epiphyses* (ends). The *metaphysis* is the flared end of the diaphysis where the shaft merges with the epiphysis. Periosteum surrounds the diaphysis; endosteum lines the medullary cavity. At the epiphyseal line, cartilage separates the epiphyses from the diaphysis.

Two types of bone tissue

Each bone consists of an outer layer of dense, smooth compact bone, which contains haversian canals, and an inner layer of spongy cancellous bone, which lacks these canals.

Cancellous bone

Cancellous bone consists of tiny spikes, called *trabeculae*, that interlace to form a latticework. Red marrow fills the spaces between the trabeculae of some bones. Cancellous bone fills the central regions of the epiphyses and the inner portions of short, flat, and irregular bones.

Compact bone

Compact bone is found in the diaphyses of long bones and the outer layers of short, flat, and irregular bones. Compact bone consists of layers of calcified matrix containing spaces occupied by osteocytes (bone cells). *Lamellae* (bone layers) are arranged around central canals (haversian canals). Small cavities called *lacunae*, which lie between the lamellae, contain osteocytes. Canaliculi (tiny canals) connect the lacunae, forming the structural units of the bone. Canaliculi also provide nutrients to bone tissue.

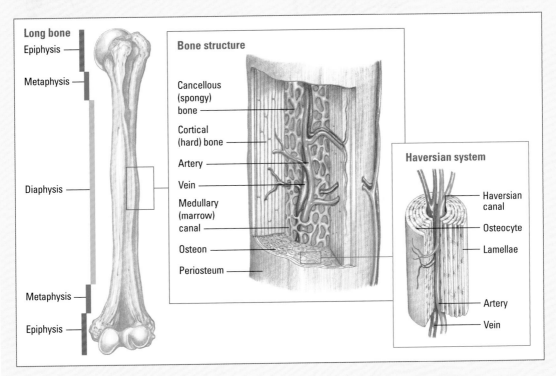

Long bone
- Epiphysis
- Metaphysis
- Diaphysis
- Metaphysis
- Epiphysis

Bone structure
- Cancellous (spongy) bone
- Cortical (hard) bone
- Artery
- Vein
- Medullary (marrow) canal
- Osteon
- Periosteum

Haversian system
- Haversian canal
- Osteocyte
- Lamellae
- Artery
- Vein

Zoom in

Bone growth and remodeling

The ossification of cartilage into bone, or *osteogenesis*, begins at about the ninth week of fetal development. The diaphyses of long bones are formed by birth, and the epiphyses begin to ossify at about that time. Here are the stages of bone growth and remodeling of the epiphyses of a long bone.

Creation of an ossification center

At about the 9th month, an ossification center develops in the epiphysis. Some cartilage cells enlarge and stimulate ossification of surrounding cells. The enlarged cells die, leaving small cavities. New cartilage continues to develop.

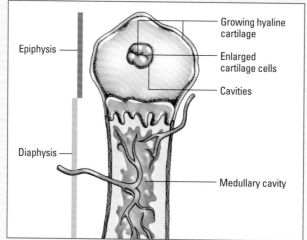

Epiphysis

Diaphysis

Growing hyaline cartilage

Enlarged cartilage cells

Cavities

Medullary cavity

Osteoblasts form bone

Osteoblasts begin to form bone on the remaining cartilage, creating the trabeculae network of cancellous bone. Cartilage continues to form on the outer surfaces of the epiphysis and along the upper surface of the epiphyseal plate.

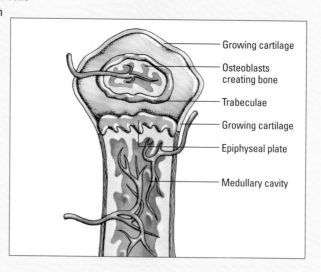

Growing cartilage

Osteoblasts creating bone

Trabeculae

Growing cartilage

Epiphyseal plate

Medullary cavity

Bone growth and remodeling *(continued)*

Bone length grows

Cartilage is replaced by compact bone near the outer surfaces of the epiphysis. Only cartilage cells on the upper surface of the epiphyseal plate continue to multiply rapidly, pushing the epiphysis away from the diaphysis. This new cartilage ossifies, creating trabeculae on the medullary side of the epiphyseal plate.

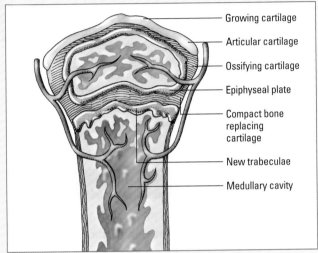

- Growing cartilage
- Articular cartilage
- Ossifying cartilage
- Epiphyseal plate
- Compact bone replacing cartilage
- New trabeculae
- Medullary cavity

Bone density begins to decrease after age 30 in women and after age 45 in men.

Remodeling

Osteoclasts produce enzymes and acids that reduce trabeculae created by the epiphyseal plate, thus enlarging the medullary cavity. In the epiphysis, osteoclasts reduce bone, making its calcium available for new osteoblasts that give the epiphysis its adult shape and proportion. In young adults, the epiphyseal plate completely ossifies (closes) and becomes the epiphyseal line; longitudinal growth of bone then ceases.

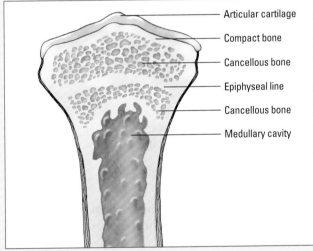

- Articular cartilage
- Compact bone
- Cancellous bone
- Epiphyseal line
- Cancellous bone
- Medullary cavity

Bone remodeling

Two types of osteocytes, osteoblasts and osteoclasts, are responsible for *remodeling*—the continuous process whereby bone is created and destroyed.

Blast them bones

Osteoblasts deposit new bone, and *osteoclasts* increase long-bone diameter. Osteoclasts promote longitudinal bone growth by reabsorbing the previously deposited bone. This growth continues until the *epiphyseal plates* ossify during late adolescence. The epiphyseal plates are cartilage that separate the *diaphysis*, or shaft of a bone, from the *epiphysis*, or end of a bone.

Cartilage

Cartilage is a dense connective tissue that consists of fibers embedded in a strong, gel-like substance. Unlike rigid bone, cartilage has the flexibility of firm plastic.

Cartilage supports and shapes various structures, such as the auditory canal, the larynx, and the intervertebral disks. It also cushions and absorbs shock, preventing direct transmission to the bone. Cartilage has no blood supply or innervation, which affects its ability to heal.

There are three types of cartilage:
1. hyaline
2. fibrous
3. elastic.

Common connector

Hyaline cartilage is the most common type of cartilage. It covers the articular bone surfaces (where one or more bones meet at a joint). It also connects the ribs to the sternum and appears in the trachea, bronchi, and nasal septum.

The strong, rigid type

Fibrous cartilage forms the symphysis pubis and the intervertebral disks. This type of cartilage is composed of small quantities of matrix and abundant fibrous elements. It's strong and rigid.

Staying flexible

Elastic cartilage, the most pliable cartilage, is located in the auditory canal, external ear, and epiglottis. Large numbers of elastic fibers give this type of cartilage elasticity and resiliency.

Joints

Joints (articulations) are points of contact between two bones that hold the bones together. Many joints also allow flexibility and movement.

Joint classification

Joints can be classified by function (extent of movement) or by structure (what they're made of). The body has three major types of joints classified by function and three major types classified by structure.

Functional classification

By function, a joint may be classified as:
1. *synarthrosis* (immovable, such as skull sutures)
2. *amphiarthrosis* (slightly movable, such as the sternocostal joint)
3. *diarthrosis* (freely movable, such as the elbow).

Structural classification

By structure, a joint may be classified as:
1. fibrous
2. cartilaginous
3. synovial.

Fibrous joints

With *fibrous joints*, the articular surfaces of the two bones are bound closely by fibrous connective tissue, and little movement is possible. Fibrous joints include *sutures, syndesmoses* (such as the radioulnar joints), and *gomphoses* (such as the dental alveolar joint).

Cartilaginous joints

With *cartilaginous joints* (also called amphiarthroses), cartilage connects one bone to another. Cartilaginous joints allow slight movement. They occur as:
- *synchondroses*, which are typically temporary joints in which the intervening hyaline cartilage converts to bone by adulthood—for example, the epiphyseal plates of long bones
- *symphyses*, which are joints with an intervening pad of fibrocartilage—for example, the symphysis pubis.

Synovial joints

The contiguous bony surfaces in the *synovial joints* are separated by a viscous, lubricating fluid—the *synovia*—and by cartilage. They're joined by ligaments lined with a synovia-producing membrane. Freely movable or diarthrosis and synovial joints include most joints of the arms and legs.

Other features of synovial joints include:

- a *joint cavity*—a potential space that separates the articulating surfaces of the two bones
- an *articular capsule*—a saclike envelope with an outer layer that's lined with a vascular synovial membrane
- *reinforcing ligaments*—fibrous tissue that connects bones within the joint and reinforces the joint capsule.

Synovial joints are freely movable and include most joints of the arms and legs.

Joint subdivisions

Based on their structure and the type of movement they allow, synovial joints fall into various subdivisions—gliding, hinge, pivot, condylar, saddle, and ball and socket.

Let it glide

Gliding joints have flat or slightly curved articular surfaces and allow gliding movements. However, because they're bound by ligaments, they may not allow movement in all directions. Examples of gliding joints are the intertarsal and intercarpal joints of the hands and feet.

Here's a hinge

With *hinge joints*, a convex portion of one bone fits into a concave portion of another. The movement of a hinge joint resembles that of a metal hinge and is limited to flexion and extension. Hinge joints include the elbow and knee.

Bones are like levers; joints are like fulcrums.

Apply your muscles and the lever moves. Wheeeee!

"Pivotal" joints

A rounded portion of one bone in a *pivot joint* fits into a groove in another bone. Pivot joints allow only uniaxial rotation of the first bone around the second. An example of a pivot joint is the head of the radius, which rotates within a groove of the ulna.

Give condylar joints a hand

With *condylar joints*, an oval surface of one bone fits into a concavity in another bone. Condylar joints allow flexion, extension, abduction, adduction, and circumduction. Examples include the radiocarpal and metacarpophalangeal joints of the hand.

Saddle up

Saddle joints resemble condylar joints but allow greater freedom of movement. The only saddle joints in the body are the carpometacarpal joints of the thumb.

Ball and socket: These joints are hip

The *ball-and-socket joint* gets its name from the way its bones connect: The spherical head of one bone fits into a concave "socket" of another bone. The body's only ball-and-socket joints are the shoulder and hip joints.

Bursae

Bursae are small synovial fluid sacs that are located at friction points around joints between tendons, ligaments, and bones.

Stress reducers

Bursae act as cushions to decrease stress on adjacent structures. Examples of bursae include the subacromial bursa (located in the shoulder) and the prepatellar bursa (located in the knee).

Quick quiz

1. Which muscle type is considered voluntary?
   A. Cardiac
   B. Smooth
   C. Skeletal
   D. Epimysium

   Answer: C. Skeletal muscle is voluntary, meaning it can be moved at will. The musculoskeletal system consists mostly of skeletal muscle.

2. Which statement about cartilage is true?
    A. It receives a generous blood supply.
    B. It protects body structures.
    C. It's completely flexible.
    D. It cushions and absorbs shock.

Answer: D. Cartilage is responsible for supporting, cushioning, and shaping body structures. Types of cartilage include fibrous, hyaline, and elastic.

3. The type of joint that permits free movement is classified as:
    A. synarthrosis.
    B. cartilaginous.
    C. diarthrosis.
    D. fibrous.

Answer: C. Diarthroses include the ankles, wrists, knees, hips, and shoulders. These joints permit free movement.

4. The carpometacarpal joints of the thumb are classified as:
    A. pivot joints.
    B. saddle joints.
    C. hinge joints.
    D. gliding joints.

Answer: B. Saddle joints are similar to condylar joints. The only saddle joints in the body are the carpometacarpal joints of the thumbs.

Scoring

☆☆☆ If you answered all four questions correctly, outstanding! You've flexed your mental muscles and are ready to tackle the next system.

☆☆ If you answered three questions correctly, splendid! You've got the bare bones of this system down pat.

☆ If you answered fewer than three questions correctly, that's okay. Let's take our osteocytes and blast through this chapter one more time.

Selected References

Hall, J. (2015). *Guyton and Hall textbook of medical physiology* (13th ed.). Philadelphia, PA: Elsevier.

Saladin, K. (2014). *Anatomy & physiology: The unity of form and function* (7th ed.). New York, NY: McGraw Hill NY.

Chapter 6

Neurosensory system

Just the facts

In this chapter, you'll learn:
- ◆ structures of the nervous system
- ◆ functions of the nervous system
- ◆ special sense organs and their functions.

A look at the neurosensory system

The nervous system coordinates all body functions, enabling a person to adapt to changes in internal and external environments. It has two main types of cells:
- neurons, the conducting cells
- neuroglia, the supportive cells.

Neuron: The basic unit

The *neuron* is the basic unit of the nervous system. This highly specialized conductor cell receives and transmits electrochemical nerve impulses. Delicate, threadlike nerve fibers called *axons* and *dendrites* extend from the central cell body and transmit signals. In a typical neuron, one axon and many dendrites extend from the cell body. (See *Parts of a neuron*, page 76.)

Axons

Axons conduct nerve impulses away from cell bodies. A typical axon has terminal branches and is wrapped in a white, fatty, segmented covering called a *myelin sheath*. The myelin sheath is produced by *Schwann cells*—phagocytic cells separated by gaps called *nodes of Ranvier*.

> I'm a neuron, the fundamental unit of the nervous system.

Zoom in

Parts of a neuron

A typical neuron, like the one shown here, has one axon and many dendrites.
A myelin sheath encloses the axon.

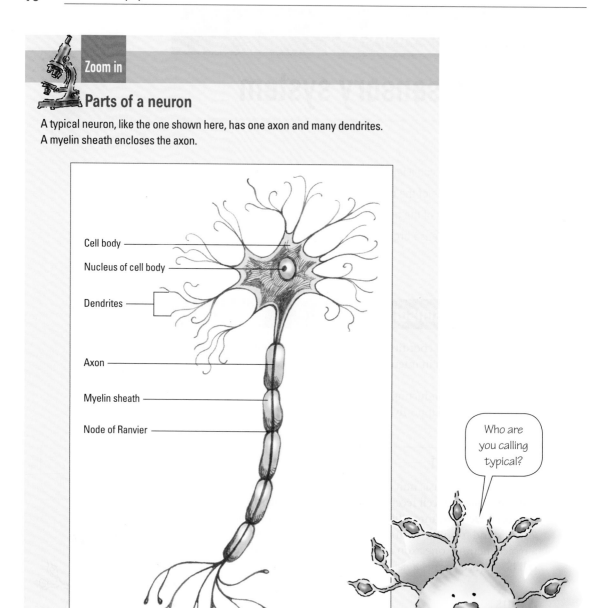

Cell body

Nucleus of cell body

Dendrites

Axon

Myelin sheath

Node of Ranvier

Who are you calling typical?

Dendrites

Dendrites are short, thick, diffusely branched extensions of the cell body that receive impulses from other cells. Dendrites conduct impulses toward the cell body.

Sending the message

Neurons are responsible for *neurotransmission*—conduction of electrochemical impulses throughout the nervous system. Neuron activity may be provoked by:

- mechanical stimuli, such as touch and pressure
- thermal stimuli, such as heat and cold
- chemical stimuli, such as external chemicals or a chemical released by the body, such as histamine. (See *How neurotransmission occurs*, page 78.)

Thermal stimuli such as heat produce neuron activity and start conduction of electrochemical impulses through the nervous system.

The reflex arc

The reflex arc—a neural relay cycle for quick motor response to a harmful sensory stimuli—requires a sensory (afferent) neuron and a motor (efferent) neuron. The stimulus triggers a sensory impulse, which travels along the dorsal root to the spinal cord. There, two synaptic transmissions occur at the same time. One synapse continues the impulse along a sensory neuron to the brain; the other immediately relays the impulse to an interneuron, which transmits it to a motor neuron. (See *The reflex arc*, page 79.)

Neuroglia

Neuroglia (also called *glial cells*) are the supportive cells of the nervous system. They form roughly 40% of the brain's bulk.

Neuroglia to know

Four types of neuroglia exist:

1. *Astroglia*, or *astrocytes*, exist throughout the nervous system. They supply nutrients to neurons and help them maintain their electrical potential. They also form part of the blood-brain barrier, which prevents harmful molecules from entering the brain.
2. *Ependymal cells* line the four small cavities in the brain, called *ventricles*, and the choroid plexuses. They help produce cerebrospinal fluid (CSF).
3. *Microglia* are phagocytic cells that ingest and digest microorganisms and waste products from injured neurons.
4. *Oligodendroglia* support and electrically insulate central nervous system (CNS) axons by forming protective myelin sheaths.

Glial is derived from the Greek word for glue. Glial cells "glue" the neurons together.

(Text continues on page 80)

How neurotransmission occurs

Neurons receive and transmit stimuli by electrochemical messages. Dendrites on the neuron receive an impulse sent by other cells and conduct it toward the cell body. The axon then conducts the impulse away from the cell.

To stimulate or inhibit

When the impulse reaches the end of the axon, it stimulates synaptic vesicles in the presynaptic axon terminal. A neurotransmitter substance is then released into the synaptic cleft between neurons. This substance diffuses across the synaptic cleft and binds to specific receptors on the postsynaptic membrane. This stimulates or inhibits stimulation of the postsynaptic neuron.

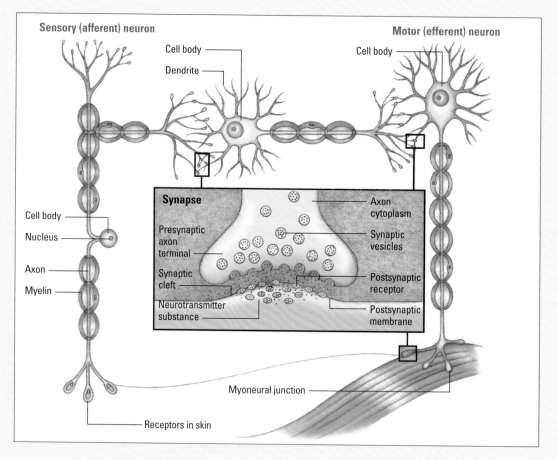

Sensory (afferent) neuron

Motor (efferent) neuron

Cell body

Dendrite

Cell body

Cell body

Nucleus

Axon

Myelin

Synapse

Presynaptic axon terminal

Synaptic cleft

Neurotransmitter substance

Axon cytoplasm

Synaptic vesicles

Postsynaptic receptor

Postsynaptic membrane

Myoneural junction

Receptors in skin

 **Now I get it!**

The reflex arc

The reflex arc is the transmission of sensory impulses to a motor neuron via the dorsal root. The motor neuron delivers the impulse to a muscle or gland, producing an immediate response.

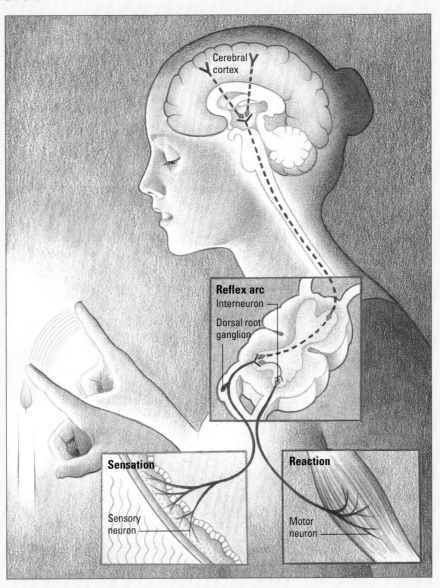

Cerebral cortex

Reflex arc
Interneuron
Dorsal root ganglion

Sensation
Sensory neuron

Reaction
Motor neuron

Central nervous system

The CNS includes the brain and the spinal cord. Encased by the bones of the skull and vertebral column, the CNS is protected by the CSF and the meninges (the dura mater, arachnoid, and pia mater).

Brain

The brain consists of the cerebrum, cerebellum, brain stem, diencephalon (thalamus and hypothalamus), limbic system, and reticular activating system.

Cerebrum

The *cerebrum* is the largest part of the brain. It houses the nerve center that controls sensory and motor activities and intelligence.

Touch of gray

The outer layer of the cerebrum, the *cerebral cortex*, consists of unmyelinated nerve fibers *(gray matter)*. The inner layer of the cerebrum consists of myelinated nerve fibers *(white matter)*.

Steady as she goes

Basal ganglia, which control motor coordination and steadiness, are found in white matter.

Bridging the hemispheres

The cerebrum has right and left hemispheres. A mass of nerve fibers known as the *corpus callosum* bridges the hemispheres, allowing communication between corresponding centers in each hemisphere. The rolling surface of the cerebrum is made up of *gyri* (convolutions) and *sulci* (creases or fissures).

The four lobes

Each cerebral hemisphere is divided into four lobes, based on anatomic landmarks and functional differences. These lobes—the frontal, temporal, parietal, and occipital—are named for the cranial bones that lie over them. (See *A close look at major brain structures*, page 82.)

Cerebellum

The *cerebellum* is the brain's second largest region. It lies behind and below the cerebrum. Like the cerebrum, it has two hemispheres. It

Basal ganglia, which control motor coordination and steadiness, are found in white matter.

also has an outer cortex of gray matter and an inner core of white matter. The cerebellum functions to maintain muscle tone, coordinate muscle movement, and control balance.

Brain stem

The *brain stem* lies immediately below the cerebrum, just in front of the cerebellum. It continues from the cerebrum above and connects with the spinal cord below.

It all stems from the brain stem

Thanks to the brain stem, I can communicate with the rest of the nervous system.

The brain stem consists of the *midbrain, pons,* and *medulla oblongata.* It relays messages between the parts of the nervous system and has three main functions:
- It produces the vital autonomic reactions necessary for survival, such as increasing heart rate and stimulating the adrenal medulla to produce epinephrine.
- It provides pathways for nerve fibers between higher and lower neural centers.
- It serves as the origin for 10 of the 12 pairs of cranial nerves (CNs).

It goes both ways

The three parts of the brain stem provide two-way conduction between the spinal cord and brain. In addition, they perform the following functions:
- The *midbrain* is the reflex center for CNs III and IV and mediates pupillary reflexes and eye movements.
- The *pons* helps regulate respirations. It connects the cerebellum with the cerebrum and links the midbrain to the medulla oblongata. It's also the reflex center for CNs V through VIII. The pons mediates chewing, taste, saliva secretion, hearing, and equilibrium.
- The *medulla oblongata* joins the spinal cord at the level of the *foramen magnum,* an opening in the occipital portion of the skull. It influences cardiac, respiratory, and vasomotor functions. It's the center for the vomiting, coughing, and hiccuping reflexes.

Diencephalon

The *diencephalon* is the part of the brain located between the cerebrum and the midbrain. It consists of the thalamus and hypothalamus, which lie deep in the cerebral hemispheres.

A close look at major brain structures

The illustration below shows the two largest structures of the brain—the cerebrum and cerebellum. Several fissures divide the cerebrum into hemispheres and lobes:

• The *fissure of Sylvius*, or the lateral sulcus, separates the temporal lobe from the frontal and parietal lobes.

• The *fissure of Rolando*, or the central sulcus, separates the frontal lobes from the parietal lobe.

• The *parieto-occipital fissure* separates the occipital lobe from the two parietal lobes.

To each lobe, a function

Each lobe has a particular function:

• The *frontal lobe* influences personality, judgment, abstract reasoning, social behavior, language expression, and voluntary movement (in the motor portion).

• The *temporal lobe* controls hearing, language comprehension, learning, understanding, and storage and recall of memories (although memories are stored throughout the entire brain).

• The *parietal lobe* interprets and integrates sensations, including pain, temperature, and touch. It also interprets size, shape, distance, vibration, and texture. The parietal lobe of the nondominant hemisphere is especially important for awareness of body shape.

• The *occipital lobe* functions mainly to interpret visual stimuli.

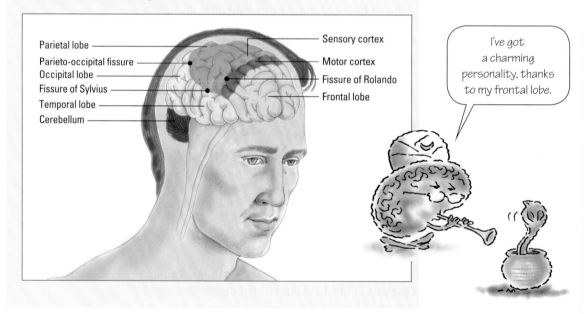

Parietal lobe
Parieto-occipital fissure
Occipital lobe
Fissure of Sylvius
Temporal lobe
Cerebellum

Sensory cortex
Motor cortex
Fissure of Rolando
Frontal lobe

I've got a charming personality, thanks to my frontal lobe.

Screening calls

The *thalamus* relays all sensory stimuli (except olfactory) as they ascend to the cerebral cortex. Its functions include primitive awareness of pain, screening of incoming stimuli, and focusing of attention.

Control center

The *hypothalamus* controls or affects body temperature, appetite, water balance, pituitary secretions, emotions, and autonomic functions, including sleeping and waking cycles.

Limbic system

The *limbic system* is a primitive brain area deep within the temporal lobe. In addition to initiating basic drives (such as hunger, aggression, and emotional and sexual arousal), the limbic system screens all sensory messages traveling to the cerebral cortex. (See *Parts of the limbic system*.)

My thalamus acts as a relay station. Sensory impulses cross synapses in the thalamus on their way to the cerebral cortex.

Zoom in

Parts of the limbic system

The illustration below shows the structures of the limbic system.

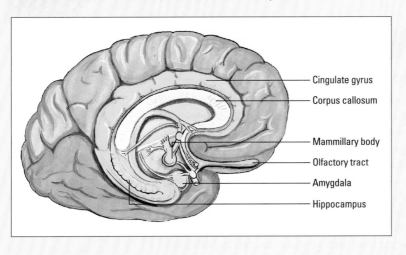

- Cingulate gyrus
- Corpus callosum
- Mammillary body
- Olfactory tract
- Amygdala
- Hippocampus

Reticular activating system

The *reticular activating system* (RAS) is a diffuse network of hyperexcitable neurons. It fans out from the brain stem through the cerebral cortex. After screening all incoming sensory information, the RAS channels it to appropriate areas of the brain for interpretation. It functions as the arousal, or alerting, system for the cerebral cortex and is crucial in maintaining consciousness. (See *Neurologic changes with aging.*)

Oxygenating the brain

Four major arteries—two vertebral and two carotid—supply the brain with oxygenated blood.

Vertebral convergence

The two *vertebral arteries* (branches of the subclavians) converge to become the basilar artery. The *basilar artery* supplies blood to the posterior brain.

Two carotids diverged in the brain...

The common carotids branch into the two internal carotids, which divide further to supply blood to the anterior brain and the middle brain. These arteries interconnect through the *circle of Willis,* an anastomosis at the base of the brain. The circle of Willis ensures that blood continually circulates to the brain despite interruption of any of the brain's major vessels. (See *Arteries of the brain.*)

Because of the circle of Willis, blood has two paths to the brain, ensuring a continuous blood supply.

Senior moment

Neurologic changes with aging

Aging affects the nervous system in many ways. For example, neurons of the central and peripheral nervous systems undergo degenerative changes. After about age 50, the number of brain cells decreases by about 1% per year. However, clinical effects usually aren't noticeable until aging advances further.

As a person ages, the hypothalamus becomes less effective at regulating body temperature. Also, the cerebral cortex undergoes a 20% neuron loss. Because nerve transmission typically slows down, elderly people may be less sensitive to or react sluggishly to external stimuli.

Changes in cerebral blood vessels also occur with aging. Blood vessels tend to narrow, and there is less growth of cerebral capillaries. In addition, for some persons, plaques and tangles develop outside of and inside neurons, although to a much lesser degree as seen with Alzheimer's disease. Damage of neurons by free radicals increases with aging, as does cerebral inflammation. All of these changes may result in a decreased ability to learn new information and retrieve information, such as being able to recall names of familiar persons.

Zoom in

Arteries of the brain

This illustration shows the inferior surface of the brain. The anterior and posterior arteries join with smaller arteries to form the circle of Willis.

Inferior view

Anterior communicating artery

Left internal carotid artery

Anterior cerebral artery

Middle cerebral artery

Posterior communicating artery

Posterior cerebral artery

Superior cerebellar artery

Basilar artery

Anterior inferior cerebellar artery

Vertebral artery

Anterior spinal artery

Posterior spinal artery

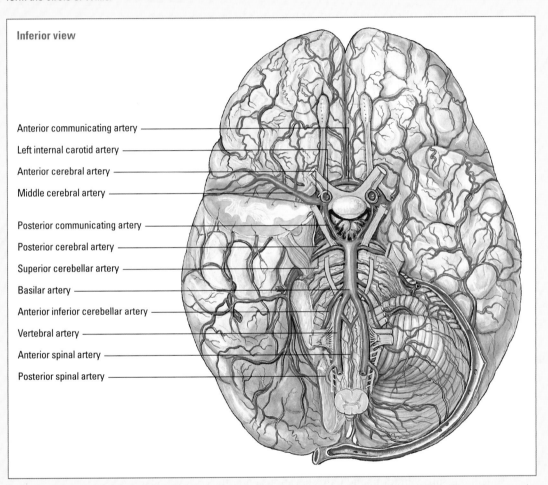

Spinal cord

The *spinal cord* is a cylindrical structure in the vertebral canal that extends from the foramen magnum at the base of the skull to the upper lumbar region of the vertebral column.

Getting on my spinal nerves

The *spinal nerves* arise from the cord. At the cord's inferior end, nerve roots cluster in the *cauda equina.*

What's the matter in the spinal cord?

Within the spinal cord, the H-shaped mass of gray matter is divided into *horns*. Horns consist mainly of neuron cell bodies. Cell bodies in the two dorsal (posterior) horns primarily relay sensations; those in the two ventral (anterior) horns play a part in voluntary and reflex motor activity. White matter surrounds the horns. This white matter consists of myelinated nerve fibers grouped in vertical columns, or *tracts*. In other words, all axons that compose one tract serve one general function, such as touch, movement, pain, and pressure. (See *A look inside the spinal cord.*)

Sensory pathways

Sensory impulses travel via the *afferent* (sensory, or ascending) *neural pathways* to the *sensory cortex* in the parietal lobe of the brain. This is where the impulses are interpreted. These impulses use two major pathways: the dorsal horn and the ganglia.

Running hot and cold

Pain and temperature sensations enter the spinal cord through the *dorsal horn*. After immediately crossing over to the opposite side of the cord, these impulses then travel to the thalamus via the spinothalamic tract.

Feeling the pressure

Touch, pressure, vibration, and pain sensations enter the cord via relay stations called *ganglia*. Ganglia are knotlike masses of nerve cell bodies on the dorsal roots of spinal nerves. Impulses travel up the cord in the dorsal column to the medulla, where they cross to the opposite side and enter the thalamus. The thalamus relays all incoming sensory impulses (except olfactory impulses) to the sensory cortex for interpretation.

Memory jogger

To remember the difference between afferent and efferent neurons, consider this:

• **Afferent** neurons cause sensation to **a**scend to the brain. (Afferent neurons are sensory and ascending.)

• **Efferent** neurons send impulses out of the brain to **e**ffect action. (Efferent neurons are motor and descending.)

Zoom in

A look inside the spinal cord

This cross section of the spinal cord shows an H-shaped mass of gray matter divided into horns, which consist primarily of neuron cell bodies. Cell bodies in the posterior, or dorsal, horn primarily relay information. Cell bodies in the anterior, or ventral, horn are needed for voluntary or reflex motor activity.

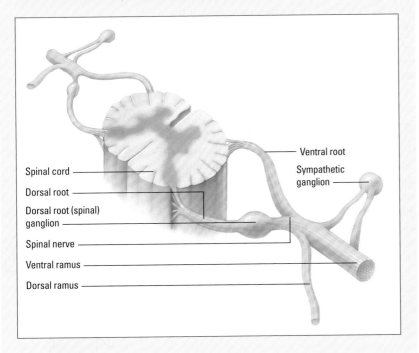

Spinal cord

Dorsal root

Dorsal root (spinal) ganglion

Spinal nerve

Ventral ramus

Dorsal ramus

Ventral root

Sympathetic ganglion

Motor pathways

Motor impulses travel from the brain to the muscles via the *efferent* (motor, or descending) *neural pathways*. Motor impulses originate in the *motor cortex* of the frontal lobe and reach the lower motor neurons of the peripheral nervous system via upper motor neurons.

Upper motor neurons originate in the brain and form two major systems:
- the pyramidal system
- the extrapyramidal system.

Fine-tuning your response

The *pyramidal system* (corticospinal tract) is responsible for fine, skilled movements of skeletal muscle. Impulses in this system travel from the motor cortex through the internal capsule to the medulla. At the medulla, they cross to the opposite side and continue down the spinal cord.

Get your motor running

The *extrapyramidal system* (extracorticospinal tract) controls gross motor movements. Impulses originate in the premotor area of the frontal lobes and travel to the pons. At the pons, the impulses cross to the opposite side. Then the impulses travel down the spinal cord to the anterior horn, where they're relayed to the lower motor neurons. These neurons, in turn, carry the impulses to the muscles. (See *Major neural pathways*.)

Reflex responses

Reflex responses occur automatically, without any brain involvement, to protect the body. Spinal nerves, which have both sensory and motor portions, mediate *deep tendon reflexes* (involuntary contractions of a muscle after brief stretching caused by tendon percussion), *superficial reflexes* (withdrawal reflexes elicited by noxious or tactile stimulation of the skin or mucous membranes), and, in infants, *primitive reflexes*.

The pyramidal system is responsible for fine, skilled movements.

Deep we go

Deep tendon reflexes include reflex responses of the biceps, triceps, brachioradialis, patella, and Achilles tendons:
- The *biceps reflex* contracts the biceps muscle and forces flexion of the forearm.
- The *triceps reflex* contracts the triceps muscle and forces extension of the forearm.
- The *brachioradialis reflex* causes supination of the hand and flexion of the forearm at the elbow.
- The *patellar reflex* forces contraction of the quadriceps muscle in the thigh with extension of the leg.
- The *Achilles reflex* forces plantar flexion of the foot at the ankle. (See *Eliciting deep tendon reflexes*, page 90.)

Rising to the superficial

Superficial reflexes are reflexes of the skin and mucous membranes. Successive attempts to stimulate these reflexes provoke increasingly limited responses. Here's a description of some superficial reflexes:
- *Plantar flexion* of the toes occurs when the lateral sole of an adult's foot is stroked from heel to great toe with a tongue blade.

Major neural pathways

Sensory and motor impulses travel through different pathways to and from the brain for interpretation.

Sensory pathways

Sensory impulses travel through two major sensory (afferent, or ascending) pathways to the sensory cortex in the cerebrum.

Motor pathways

Motor impulses travel from the motor cortex in the cerebrum to the muscles via motor (efferent, or descending) pathways.

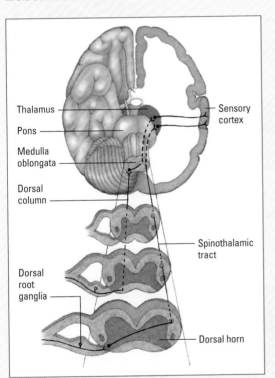

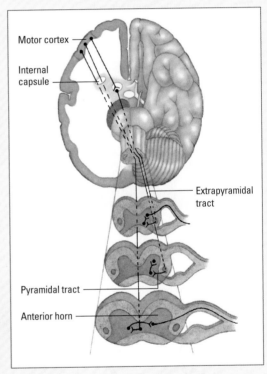

- *Babinski's reflex* is an upward movement of the great toe and fanning of the little toes that occurs in children under age 2 in response to stimulation of the outer margin of the sole of the foot. This reflex is an abnormal finding in adults.
- In men, the *cremasteric reflex* is stimulated by stroking the inner thigh. This forces the contraction of the cremaster muscle and elevation of the testicle on the side of the stimulus.

Eliciting deep tendon reflexes

There are five deep tendon reflexes. Methods for eliciting these reflexes are described below.

Biceps reflex

Placing the thumb or index finger over the biceps tendon and the remaining fingers loosely over the triceps muscle, strike the thumb or index finger over the biceps tendon with the pointed end of the reflex hammer. Watch and feel for the contraction of the biceps muscle and flexion of the forearm.

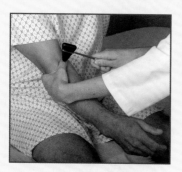

Triceps reflex

Strike the triceps tendon about 2" (5 cm) above the olecranon process on the extensor surface of the upper arm. Watch for contraction of the triceps muscle and extension of the forearm.

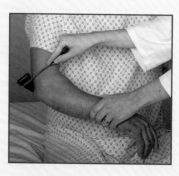

Brachioradialis reflex

Strike the radius about 1" to 2" (2.5 to 5 cm) above the wrist and watch for supination of the hand and flexion of the forearm at the elbow.

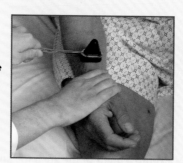

Patellar reflex

Strike the patellar tendon just below the patella and look for contraction of the quadriceps muscle in the thigh with extension of the leg.

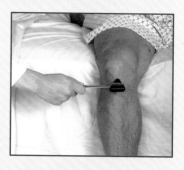

Achilles reflex

With the foot flexed and supporting the plantar surface, strike the Achilles tendon. Watch for plantar flexion of the foot at the ankle.

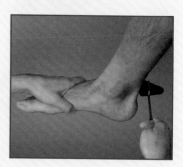

> Deep tendon reflexes help protect the body.

- The *abdominal reflexes* are induced by stroking the sides of the abdomen above and below the umbilicus, moving from the periphery toward the midline. Movement of the umbilicus toward the stimulus is normal.

Let's get primitive

Primitive reflexes are abnormal in adults but normal in infants, whose central nervous systems are immature. As the neurologic system matures, these reflexes disappear. The primitive reflexes are *grasping*, *sucking*, and *glabella*:
- The application of gentle pressure to an infant's palm results in grasping.
- An infantile sucking reflex to ingest milk is a primitive response to oral stimuli.
- The glabella reflex is elicited by repeatedly tapping the infant on the bridge of the nose or between the eyebrows. The normal response is persistent blinking.

Protective structures

The brain and spinal cord are protected from shock and infection by the bony skull and vertebrae, CSF, and three membranes: the dura mater, arachnoid membrane, and pia mater.

Hey, we all need a little protection sometimes!

Dura mater

The *dura mater* is tough, fibrous, leatherlike tissue composed of two layers—the endosteal dura and meningeal dura.

The unending endosteal dura

The *endosteal dura* forms the periosteum of the skull and is continuous with the lining of the vertebral canal.

The durable meningeal dura

The *meningeal dura* is a thick membrane that covers the brain, dipping between the brain tissue and providing support and protection.

Arachnoid membrane

The *arachnoid membrane* is a thin, fibrous membrane that hugs the brain and spinal cord, though not as precisely as the pia mater.

Pia mater

The *pia mater* is a continuous, delicate layer of connective tissue that covers and contours the spinal tissue and brain.

The spaces between

The *subdural space* lies between the dura mater and the arachnoid membrane. The *subarachnoid space* lies between the arachnoid membrane and the pia mater. Within the subarachnoid space and the brain's four ventricles is CSF, a fluid composed of water and traces of organic materials (especially protein), glucose, and electrolytes. This fluid protects the brain and spinal tissue from jolts and blows.

Peripheral nervous system

The peripheral nervous system consists of the cranial nerves, spinal nerves, and autonomic nervous system (ANS).

Cranial nerves

Twelve pairs of cranial nerves transmit motor or sensory messages (or both) primarily between the brain or brain stem and the head and neck. All cranial nerves except the olfactory and optic nerves exit from the midbrain, pons, or medulla oblongata of the brain stem. (See *Exit points for the cranial nerves.*)

Each spinal nerve gets its name from the vertebra immediately *below* its exit point from the spinal cord.

Spinal nerves

Each of the 31 pairs of spinal nerves is named for the vertebra immediately below the nerve's exit point from the spinal cord. From top to bottom, they're designated as C1 through S5 and the coccygeal nerve. Each spinal nerve consists of afferent (sensory) and efferent (motor) neurons, which carry messages to and from particular body regions, called *dermatomes.* (See *The spinal nerves,* page 94.)

Autonomic nervous system

The vast ANS *innervates* (supplies nerves to) all internal organs. Sometimes known as *visceral efferent nerves,* the nerves of the ANS carry messages to the viscera from the brain stem and neuroendocrine regulatory centers. The ANS has two major subdivisions: the

Zoom in

Exit points for the cranial nerves

As this illustration reveals, 10 of the 12 pairs of cranial nerves (CNs) exit from the brain stem. The remaining two pairs—the olfactory and optic nerves—exit from the forebrain.

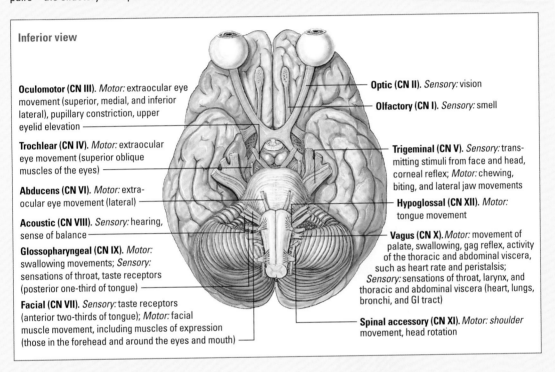

Inferior view

Oculomotor (CN III). *Motor:* extraocular eye movement (superior, medial, and inferior lateral), pupillary constriction, upper eyelid elevation

Trochlear (CN IV). *Motor:* extraocular eye movement (superior oblique muscles of the eyes)

Abducens (CN VI). *Motor:* extraocular eye movement (lateral)

Acoustic (CN VIII). *Sensory:* hearing, sense of balance

Glossopharyngeal (CN IX). *Motor:* swallowing movements; *Sensory:* sensations of throat, taste receptors (posterior one-third of tongue)

Facial (CN VII). *Sensory:* taste receptors (anterior two-thirds of tongue); *Motor:* facial muscle movement, including muscles of expression (those in the forehead and around the eyes and mouth)

Optic (CN II). *Sensory:* vision

Olfactory (CN I). *Sensory:* smell

Trigeminal (CN V). *Sensory:* transmitting stimuli from face and head, corneal reflex; *Motor:* chewing, biting, and lateral jaw movements

Hypoglossal (CN XII). *Motor:* tongue movement

Vagus (CN X). *Motor:* movement of palate, swallowing, gag reflex, activity of the thoracic and abdominal viscera, such as heart rate and peristalsis; *Sensory:* sensations of throat, larynx, and thoracic and abdominal viscera (heart, lungs, bronchi, and GI tract)

Spinal accessory (CN XI). *Motor:* shoulder movement, head rotation

sympathetic (thoracolumbar) nervous system and *parasympathetic* (craniosacral) nervous system.

When one system stimulates certain smooth muscles to contract or a gland to secrete, the other system inhibits that action. Through this dual innervation, the two divisions counterbalance each other's activities to keep body systems running smoothly.

Sympathetic nervous system

Sympathetic nerves called *preganglionic neurons* exit the spinal cord between the levels of the first thoracic and second lumbar vertebrae.

Body shop

The spinal nerves

There are 31 pairs of spinal nerves. After leaving the spinal cord, many nerves join together to form networks called *plexuses*, as this illustration shows.

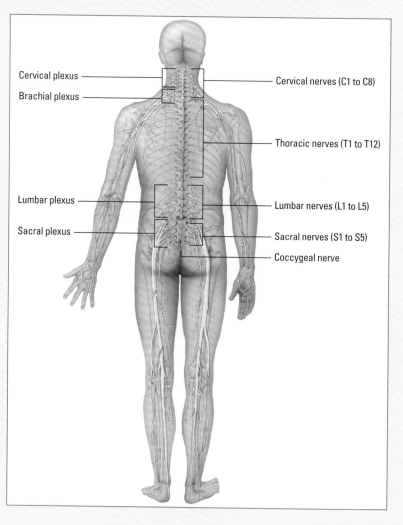

Cervical plexus

Brachial plexus

Cervical nerves (C1 to C8)

Thoracic nerves (T1 to T12)

Lumbar plexus

Lumbar nerves (L1 to L5)

Sacral plexus

Sacral nerves (S1 to S5)

Coccygeal nerve

Ganglia branch out

When they leave the spinal cord, preganglionic neurons enter small ganglia near the cord. The ganglia form a chain that spreads the impulse to *postganglionic neurons.* Postganglionic neurons reach many organs and glands and can produce widespread, generalized physiologic responses. These responses include:
- vasoconstriction
- elevated blood pressure
- enhanced blood flow to skeletal muscles
- increased heart rate and contractility
- increased respiratory rate
- smooth muscle relaxation of the bronchioles, GI tract, and urinary tract
- sphincter contraction
- pupillary dilation and ciliary muscle relaxation
- increased sweat gland secretion
- reduced pancreatic secretion.

My postganglionic neurons are making my blood pressure rise!

Parasympathetic nervous system

Fibers of the parasympathetic nervous system leave the CNS by way of the cranial nerves from the midbrain and medulla and the spinal nerves between the second and fourth sacral vertebrae (S2 to S4).

See ya later, CNS

After leaving the CNS, the long preganglionic fiber of each parasympathetic nerve travels to a ganglion near a particular organ or gland. The short postganglionic fiber enters the organ or gland. This creates a more specific response involving only one organ or gland.

Such a response might be:
- reduction in heart rate, contractility, and conduction velocity
- bronchial smooth muscle constriction
- increased GI tract tone and peristalsis, with sphincter relaxation

Faster! Get with the program!

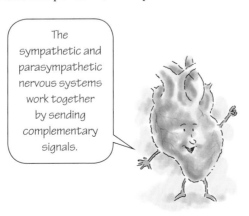

The sympathetic and parasympathetic nervous systems work together by sending complementary signals.

You can take it easy now.

- increased bladder tone and urinary system sphincter relaxation
- vasodilation of external genitalia, causing erection
- pupil constriction
- increased pancreatic, salivary, and lacrimal secretions.

Special sense organs

Sensory stimulation allows the body to interact with the environment. The distal ends of the dendrites of sensory neurons serve as sensory receptors, sending messages to the brain. The brain also receives stimulation from the special sense organs—the eyes, ears, and gustatory and olfactory organs.

Eye

The *eyes* are the organs of vision. They contain about 70% of the body's sensory receptors. Although each eye measures about 1′ (2.5 cm) in diameter, only its anterior surface is visible. Extraocular and intraocular eye structures work together for proper eye function.

The eyelids protect the eyes, regulate the entrance of light, and distribute tears over the eyes.

Extraocular eye structures

Extraocular muscles hold the eyes in place and control their movement. Their coordinated action keeps both eyes parallel and creates binocular vision. These muscles have mutually antagonistic actions: As one muscle contracts, its opposing muscle relaxes.

Extraocular structures include the eyelids, conjunctivae, and lacrimal apparatus. Together with the extraocular muscles, these structures support and protect the eyeball.

Eyelids
The *eyelids* (also called the *palpebrae*) are loose folds of skin that cover the anterior portion of the eye. The lid margins contain hair follicles, which contain eyelashes and sebaceous glands.

In the blink of an eye
When open, the upper eyelid extends beyond the *limbus* (the junction of the cornea and the sclera) and covers a small portion of the iris.
The eyelids contain three types of glands:
- *meibomian glands*—sebaceous glands that secrete sebum, an oily substance that keeps the eye lubricated
- *glands of Zeis*—modified sebaceous glands connected to the follicles of the eyelashes
- *Moll's glands*—ordinary sweat glands.
When closed, the upper and lower eyelids cover the eye completely.

Conjunctivae

Conjunctivae are thin mucous membranes that line the inner surface of each eyelid and the anterior portion of the sclera. Conjunctivae guard the eye from invasion by foreign matter. The *palpebral conjunctiva*—the portion that lines the inner surface of the eyelids—appears shiny pink or red. The *bulbar* (or ocular) *conjunctiva*, which joins the palpebral portion and covers the exposed part of the sclera, contains many small, normally visible blood vessels.

We just left the nasolacrimal duct. Last stop, the nose.

Lacrimal apparatus

The structures of the *lacrimal apparatus* (lacrimal glands, punctum, lacrimal sac, and nasolacrimal duct) lubricate and protect the cornea and conjunctivae by producing and absorbing tears. Tears keep the cornea and conjunctivae moist. Tears also contain *lysozyme*, an enzyme that protects against bacterial invasion.

Cry me a river

As the eyelids blink, they direct the flow of tears from the lacrimal ducts to the *inner canthus*, the medial angle between the eyelids. Tears pool at the inner canthus and drain through the *punctum*, a tiny opening. From there, they flow through the *lacrimal canals* into the lacrimal sac. Lastly, they drain through the nasolacrimal duct and into the nose. (See *A close look at tears*, page 98.)

Intraocular eye structures

Intraocular structures within the eyeball are directly involved with vision. (See *Looking at intraocular structures*, page 99.)

Anterior segment

The sclera, cornea, iris, pupil, anterior chamber, aqueous humor, lens, ciliary body, and posterior chamber are found in the anterior segment.

Moisture plays an important role in the eye. For example, tears protect the cornea by keeping it moist.

Sclera and cornea

The white *sclera* coats four-fifths of the outside of the eyeball, maintaining its size and form. The *cornea* is continuous with the sclera at the limbus, revealing the pupil and iris. A smooth, transparent tissue, the cornea has no blood supply. The corneal epithelium merges with the bulbar conjunctiva at the limbus. The cornea is highly sensitive to touch and is kept moist by tears.

Iris and pupil

The *iris* is a circular contractile disk that contains smooth and radial muscles. It has an opening in the center for the *pupil*. Eye

Zoom in

A close look at tears

Tears begin in the lacrimal gland and drain through the nasolacrimal duct into the nose.

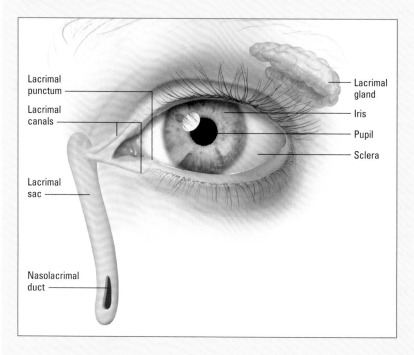

color depends on the amount of pigment in the endothelial layers of the iris. Pupil size is controlled by involuntary dilatory and sphincter muscles in the posterior region of the iris that regulate light entry.

Anterior chamber and aqueous humor

The *anterior chamber* is a cavity bounded in front by the cornea and behind by the lens and iris. It's filled with a clear, watery fluid called *aqueous humor*.

Lens

The *lens* is situated directly behind the iris at the pupillary opening. Composed of transparent fibers in an elastic membrane called the *lens capsule*, the lens acts like a camera lens, refracting and focusing light onto the retina.

Zoom in

Looking at intraocular structures

Some intraocular structures, such as the sclera, cornea, iris, pupil, and anterior chamber, are visible to the naked eye. Others, such as the retina, are visible only with an ophthalmoscope. These illustrations show the major structures within the eye.

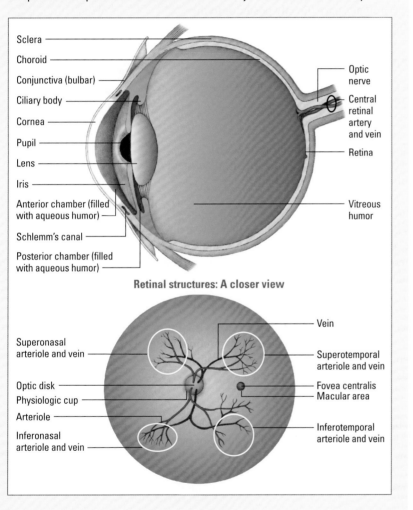

Sclera

Choroid

Conjunctiva (bulbar)

Ciliary body

Cornea

Pupil

Lens

Iris

Anterior chamber (filled with aqueous humor)

Schlemm's canal

Posterior chamber (filled with aqueous humor)

Optic nerve

Central retinal artery and vein

Retina

Vitreous humor

Retinal structures: A closer view

Superonasal arteriole and vein

Optic disk

Physiologic cup

Arteriole

Inferonasal arteriole and vein

Vein

Superotemporal arteriole and vein

Fovea centralis

Macular area

Inferotemporal arteriole and vein

Ciliary body

The *ciliary body* (three muscles along with the iris that make up the anterior part of the vascular uveal tract) controls the lens thickness. Together with the coordinated action of muscles in the iris, the ciliary body regulates the light focused through the lens onto the retina.

Posterior chamber

The *posterior chamber* is a small space directly posterior to the iris but anterior to the lens. It's filled with aqueous humor.

Posterior segment

The vitreous humor, posterior sclera, choroid, and retina are found in the posterior segment.

The posterior segment of the eye has three layers: fibrous, vascular, and sensory.

Vitreous humor

The *vitreous humor* consists of a thick, gelatinous material that fills the space behind the lens. There, it maintains placement of the retina and the spherical shape of the eyeball.

Posterior sclera and choroid

The *posterior sclera* is a white, opaque, fibrous layer that covers the posterior segment of the eyeball. It continues back to the dural sheath, covering the optic nerve. The *choroid* lies beneath the posterior sclera. It contains many small arteries and veins.

Retina

The *retina* is the innermost coat of the eyeball. It receives visual stimuli and sends them to the brain. Each of the four sets of retinal vessels contains a transparent arteriole and vein as well as the optic disk, the physiologic cup, rods and cones, and the macula.

The optimal optic disk

Arterioles and veins become progressively thinner as they leave the optic disk. The *optic disk* is a well-defined, 1.5-mm round or oval area on the retina. Creamy yellow to pink in color, the optic disk allows the optic nerve to enter the retina at a point called the *nerve head*. A whitish to grayish crescent of scleral tissue may be present on the lateral side of the disk.

The cup within the disk

The *physiologic cup* is a light-colored depression within the optic disk on the temporal side. It covers one-third of the center of the disk.

Visionaries

Photoreceptor neurons called *rods* and *cones* compose the visual receptors of the retina. These receptors are responsible for vision.

Count macula

The *macula* is lateral to the optic disk. It's slightly darker than the rest of the retina and without visible retinal vessels. A slight depression in the center of the macula, known as the *fovea centralis*, contains the heaviest concentration of cones and is a main receptor for vision and color.

The optic disk has no light receptors and is therefore a "blind spot." But I compensate so that there's usually no apparent gap in what is seen.

Vision pathway

Intraocular structures perceive and form images and then send them to the brain for interpretation. To interpret these images properly, the brain relies on structures along the vision pathway. The *vision pathway* uses the optic nerve, optic chiasm, and retina to create the proper visual fields.

Crisscrossing tracts

In the *optic chiasm*, fibers from the nasal aspects of both retinas cross to the opposite sides, and fibers from the temporal portions remain uncrossed. These crossed and uncrossed fibers form the *optic tracts*. Injury to one of the optic nerves can cause blindness in the corresponding eye. An injury or lesion in the optic chiasm can cause partial vision loss (for example, loss of the two temporal visual fields).

Focusing on the fovea centralis

Image formation begins when eye structures refract light rays from an object. Normally, the cornea, aqueous humor, lens, and vitreous humor refract light rays from an object, focusing them on the fovea centralis, where an inverted and reversed image clearly forms. Within the retina, rods and cones turn the projected image into an impulse and transmit it to the optic nerve.

I see! We turn projected images into impulses that we then transmit to the optic nerve.

Follow the tracts to the cerebral cortex

The impulse travels to the optic chiasm (where the two optic nerves unite and split again into two optic tracts) and then continues into the optic section of the cerebral cortex. There, the inverted and reversed image on the retina is processed by the brain to create an image as it truly appears in the field of vision.

Ear

The *ears* are the organs of hearing. They also maintain the body's equilibrium. The ear is divided into three main parts: external, middle, and inner. (See *Ear structures*.)

External ear structures

The *external ear* consists of the auricle and the external auditory canal. The *mastoid process* isn't part of the external ear but is an important bony landmark behind the lower part of the auricle.

Zoom in

Ear structures

The ear is the organ of hearing. The structures of its three sections—external, middle, and inner—are illustrated below.

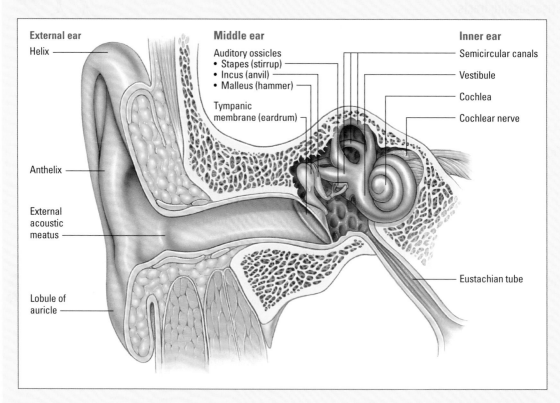

External ear
Helix

Anthelix

External acoustic meatus

Lobule of auricle

Middle ear
Auditory ossicles
• Stapes (stirrup)
• Incus (anvil)
• Malleus (hammer)

Tympanic membrane (eardrum)

Inner ear
Semicircular canals

Vestibule

Cochlea

Cochlear nerve

Eustachian tube

Auricle

The *auricle* (pinna) is the outer, visual protrusion of the ear. It helps collect and direct incoming sound into the external auditory canal.

External auditory canal

The *external auditory canal* is a narrow chamber that connects the auricle with the tympanic membrane. This canal transmits sound to the eardrum and tympanic membrane.

Middle ear structures

The *middle ear* is also called the *tympanic cavity*. It's an air-filled cavity within the hard portion of the temporal bone. The tympanic cavity is lined with mucosa. It's bound distally by the tympanic membrane and medially by the oval and round windows. The eustachian tube equalizes pressure within the ear and the small bones of the middle ear conduct vibration.

Tympanic membrane

The *tympanic membrane* consists of layers of skin, fibrous tissue, and a mucous membrane. It transmits sound vibrations to the internal ear.

Eustachian tube

The *eustachian*, or auditory, *tube* extends downward, forward, and inward from the middle ear cavity to the nasopharynx. It has a useful function: It allows the pressure against inner and outer surfaces of the tympanic membrane to equalize, preventing rupture and allowing for proper transfer of sound waves.

Oval window

The *oval window* (fenestra ovalis) is an opening in the wall between the middle and inner ears into which part of the *stapes* (a tiny bone of the middle ear) fits. It transmits vibrations to the inner ear.

Round window

The *round window* (fenestra cochleae) is another opening in the same wall. It's enclosed by the secondary tympanic membrane. Like the oval window, the round window transmits vibrations to the inner ear.

Small bones

The middle ear contains three small bones, called *ossicles*, that conduct vibratory motion of the tympanum to the oval window. The ossicles are:

- the *malleus* (hammer), which attaches to the tympanic membrane and transfers sound to the incus

- the *incus* (anvil), which articulates the malleus and the stapes and carries vibration to the stapes
- the *stapes* (stirrup), which connects vibratory motion from the incus to the oval window.

Inner ear structures

In the inner ear, vibration excites receptor nerve endings. A bony labyrinth and a membranous labyrinth combine to form the inner ear. The inner ear contains the vestibule, cochlea, and semicircular canals.

Vestibule

The *vestibule* is located posterior to the cochlea and anterior to the semicircular canals. It serves as the entrance to the inner ear. It houses two membranous sacs, the *saccule* and *utricle*. Suspended in a fluid called *perilymph*, the saccule and utricle sense gravity changes and linear and angular acceleration.

Cochlea

The *cochlea*, a bony, spiraling cone, extends from the anterior part of the vestibule. Within it lies the *cochlear duct*, a triangular, membranous structure that houses the *organ of Corti*. The receptor organ for hearing, the organ of Corti, transmits sound to the cochlear branch of the acoustic nerve (CN VIII).

Semicircular canals

The three *semicircular canals* project from the posterior aspect of the vestibule. Each canal is oriented in one of three planes: superior, posterior, and lateral. The *semicircular duct* traverses the canals and connects with the utricle anteriorly. The *crista ampullaris* sits at the end of each canal and contains hair cells and support cells. It's stimulated by sudden movements or changes in the rate or direction of movement.

Hearing pathways

For hearing to occur, sound waves travel through the ear by two pathways—air conduction and bone conduction:
- *Air conduction* occurs when sound waves travel in the air through the external and middle ear to the inner ear.
- *Bone conduction* occurs when sound waves travel through bone to the inner ear. (See *Sound transmission*.)

For hearing to occur, sound waves travel through the ear by two pathways—air conduction and bone conduction.

Interpreting the vibrations, man

Vibrations transmitted through air and bone stimulate nerve impulses in the inner ear. The cochlear branch of the acoustic nerve transmits these vibrations to the auditory area of the cerebral cortex. The cerebral cortex then interprets the sound.

Now I get it!

Sound transmission

The illustration below details the transmission of sound through the structures of the middle and inner ear.

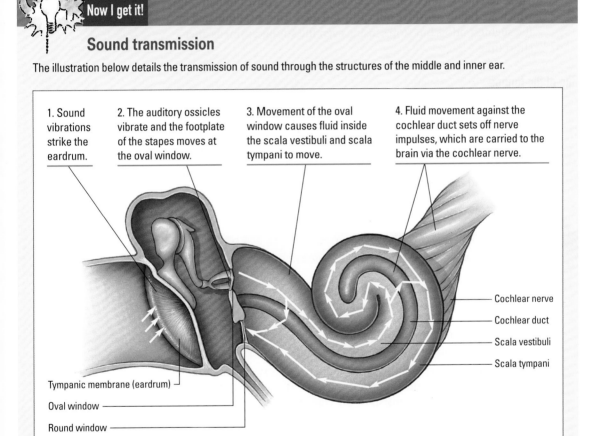

1. Sound vibrations strike the eardrum.

2. The auditory ossicles vibrate and the footplate of the stapes moves at the oval window.

3. Movement of the oval window causes fluid inside the scala vestibuli and scala tympani to move.

4. Fluid movement against the cochlear duct sets off nerve impulses, which are carried to the brain via the cochlear nerve.

Cochlear nerve
Cochlear duct
Scala vestibuli
Scala tympani

Tympanic membrane (eardrum)
Oval window
Round window

Nose and mouth

The *nose* is the sense organ for smell. The mucosal epithelium that lines the uppermost portion of the nasal cavity houses receptors for fibers of the olfactory nerve (CN I).

Good old olfactory

These receptors, called *olfactory* (smell) *receptors,* consist of hair cells, which are highly sensitive but easily fatigued. They're stimulated by the slightest odors but stop sensing even the strongest smells after a short time.

Slip of the tongue

The tongue and the roof of the mouth contain most of the receptors for the taste nerve fibers (located in branches of CNs VII and IX). Called *taste buds,* these receptors are stimulated by chemicals. They respond to four taste sensations: sweet, sour, bitter, and salty. All the other flavors a person senses result from a combination of olfactory-receptor and taste-bud stimulation.

Quick quiz

1. The components of the CNS include:
    A. the spinal cord and cranial nerves.
    B. the brain and spinal cord.
    C. the sympathetic and parasympathetic nervous systems.
    D. the cranial nerves and spinal nerves.

Answer: B. The two main divisions of the nervous system are the CNS, which includes the brain and spinal cord, and the peripheral nervous system, which consists of the cranial nerves, spinal nerves, and ANS.

2. The brain is protected from shock and infection by:
    A. bones, the meninges, and CSF.
    B. gray matter, bones, and the primitive structures.
    C. the blood-brain barrier, CSF, and white matter.
    D. axons, neurons, and meninges.

Answer: A. Bones (the skull and vertebral column), the meninges, and CSF protect the brain from shock and infection.

3. The visual receptors of the retina are composed of:
   A. the pupil and lens.
   B. the optic disk and optic nerve.
   C. vitreous humor and aqueous humor.
   D. rods and cones.

Answer: D. Photoreceptor neurons called rods and cones compose the visual receptors of the retina.

4. The external ear consists of:
   A. the vestibule, cochlea, and semicircular ducts.
   B. the tympanic membrane, oval window, and round window.
   C. the auricle and external auditory canal.
   D. the malleus, incus, and stapes.

Answer: C. The auricle and external auditory canal are part of the external ear.

5. The cranial nerves transmit motor and sensory messages between the:
   A. spine and body dermatomes.
   B. brain and the head and neck.
   C. viscera and the brain.
   D. brain and skeletal muscles.

Answer: B. The 12 pairs of cranial nerves transmit motor (efferent) and sensory (afferent) messages between the brain or brain stem and the head and neck.

Scoring

☆☆☆ If you answered all five questions correctly, stupendous! You're definitely brainy.

☆☆ If you answered four questions correctly, remarkable. Your gray matter is working at lightning speed.

☆ If you answered fewer than four questions correctly, don't get nervous. Give your brain another workout by reviewing the chapter again.

Just for fun!

The ANS has two major subdivisions. Unscramble the words in the boxes on the left to reveal the names of these two nervous systems. Then draw a line from the boxes on the left to the list on the right, linking each system to its physiologic responses.

CAMPSITETHY

— — — — — — — — — —

APHARMACYTESTPI

— — — — — — — — — —

A. Bronchial smooth muscle constriction
B. Vasoconstriction
C. Reduction of heart rate, contractility, and conduction velocity
D. Elevated blood pressure
E. Increased heart rate and contractility
F. Increased GI tract tone and peristalsis, with sphincter relaxation
G. Enhanced blood flow to skeletal muscles
H. Increased respiratory rate
I. Vasodilation of external genitalia, causing erection
J. Increased pancreatic, salivary, and lacrimal secretions
K. Smooth muscle relaxation of the bronchioles, GI tract, and urinary tract
L. Sphincter contraction
M. Pupil constriction
N. Pupillary dilation and ciliary muscle relaxation
O. Increased sweat gland secretion
P. Reduced pancreatic secretion

Answer: SYMPATHETIC: B, D, E, G, H, K, L, N, O, P; PARASYMPATHETIC: A, C, F, I, J, M

Selected References

McCance, K. L., & Huether, S. E. (2015). *Pathophysiology: The biologic basis for disease in adults and children* (7th ed.). Elsevier Health Sciences. St. Louis. https://books.google.com/books. ISBN No.: 0323293751.

National Institute on Aging. (January 22, 2015). *The changing brain in healthy aging.* https://www.nia.nih.gov/

Chapter 7

Endocrine system

Just the facts

In this chapter, you'll learn:

◆ the functions of endocrine glands

◆ hormone release and transportation in the endocrine system

◆ the role of receptors in the influence of hormones on cells.

A look at the endocrine system

The three major components of the endocrine system are:

- *glands*—specialized cell clusters or organs
- *hormones*—chemical substances secreted by glands in response to stimulation
- *receptors*—protein molecules that bind specifically with other molecules, such as hormones, to trigger specific physiologic changes in a target cell.

Along with the nervous system, the endocrine system regulates and integrates the body's metabolic activities.

Glands

The major glands of the endocrine system are:

- pituitary gland
- thyroid gland
- parathyroid glands
- adrenal glands
- pancreas
- thymus
- pineal gland
- gonads (ovaries and testes). (See *Components of the endocrine system*, page 110.)

Components of the endocrine system

Endocrine glands secrete hormones directly into the bloodstream to regulate body function. This illustration shows the location of the major endocrine glands.

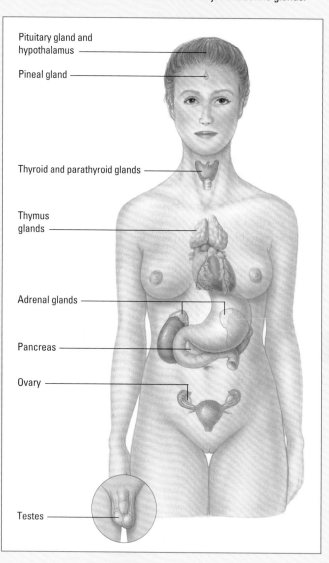

Pituitary gland and hypothalamus

Pineal gland

Thyroid and parathyroid glands

Thymus glands

Adrenal glands

Pancreas

Ovary

Testes

The pituitary gland

The *pituitary gland* (also called the *hypophysis* or *master gland*) rests in the *sella turcica,* a depression in the sphenoid bone at the base of the brain. Blood vessels and nerves carry messages from the hypothalamus to the pea-sized pituitary gland through the infundibulum. (See *How the hypothalamus affects endocrine activities,* page 112.) The pituitary gland has two main regions: the anterior pituitary and the posterior pituitary.

Anterior pituitary

The *anterior pituitary* (adenohypophysis) is the larger region of the pituitary gland. It produces at least six hormones:
- growth hormone (GH), or somatotropin
- thyroid-stimulating hormone (TSH), or thyrotropin
- corticotropin
- follicle-stimulating hormone (FSH)
- luteinizing hormone (LH)
- prolactin.

Posterior pituitary

The *posterior pituitary* makes up about 25% of the gland. It serves as a storage area for antidiuretic hormone (ADH), also known as *vasopressin,* and oxytocin, which are produced by the hypothalamus.

Thyroid gland

The *thyroid* lies directly below the larynx, partially in front of the trachea. Its two lateral lobes—one on either side of the trachea—join with a narrow tissue bridge, called the *isthmus,* to give the gland its butterfly shape.

Two lobes that function as one

The two lobes of the thyroid function as one unit to produce the hormones *triiodothyronine (T_3), thyroxine (T_4),* and *calcitonin.* (See *Thyroid stimulation,* page 113.)

T_3 and T_4 equal thyroid hormone

T_3 and T_4 are collectively referred to as *thyroid hormone.* The body's major metabolic hormone, thyroid hormone, regulates metabolism by speeding cellular respiration. The effects can be seen on body temperature, heart rate, brain maturation, growth, higher levels of attention, and quicker reflexes.

Now I get it!

How the hypothalamus affects endocrine activities

Hypothalamus and pituitary

Anterior and posterior pituitary secretions are controlled by signals from the hypothalamus:

• As shown on the left side of the illustration below, the hypothalamic neuron produces antidiuretic hormone (ADH), which travels down the axon and is stored in secretory granules in nerve endings in the posterior pituitary for later release.

• As shown on the right side of the illustration below, the hypothalamus stimulates the anterior pituitary's production of its many hormones. A hypothalamic neuron manufactures inhibitory and stimulatory hormones and secretes them into a capillary of the portal system. The hormones travel down the pituitary stalk to the anterior pituitary. There, they cause inhibition or release of many pituitary hormones, including corticotropin, thyroid-stimulating hormone, growth hormone, follicle-stimulating hormone, luteinizing hormone, and prolactin.

The hypothalamus is the integrative center for the endocrine and autonomic nervous systems.

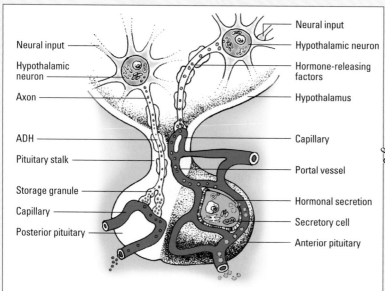

Neural input

Hypothalamic neuron

Axon

ADH

Pituitary stalk

Storage granule

Capillary

Posterior pituitary

Neural input

Hypothalamic neuron

Hormone-releasing factors

Hypothalamus

Capillary

Portal vessel

Hormonal secretion

Secretory cell

Anterior pituitary

Now I get it!

Thyroid stimulation

Thyroid cells store a hormone precursor, colloidal iodinated thyroglobulin, which contains iodine and thyroglobulin. When stimulated by thyroid-stimulating hormone (TSH), a follicular cell (shown below) takes up some stored thyroglobulin by *endocytosis*—the reverse of exocytosis.

The cell membrane extends fingerlike projections into the colloid, and then pulls portions of it back into the cell. Lysosomes fuse with the colloid, which is then degraded into triiodothyronine (T_3) and thyroxine (T_4). These thyroid hormones are released into the circulation and lymphatic system by exocytosis.

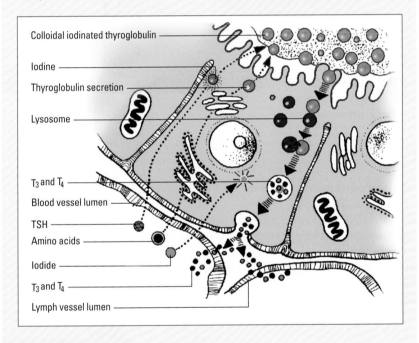

Colloidal iodinated thyroglobulin

Iodine

Thyroglobulin secretion

Lysosome

T_3 and T_4

Blood vessel lumen

TSH

Amino acids

Iodide

T_3 and T_4

Lymph vessel lumen

Calcium balancing act

Calcitonin maintains the calcium level of blood. It does this by inhibiting the release of calcium from bone. Secretion of calcitonin is controlled by the calcium concentration of the fluid surrounding the thyroid cells.

Parathyroid glands

The *parathyroid glands* are the body's smallest known endocrine glands. These glands are embedded on the posterior surface of the thyroid, one in each corner.

Calcium cohort

Working together as a single gland, the parathyroid glands produce *parathyroid hormone* (PTH). The main function of PTH is to help regulate the blood's calcium balance. This hormone adjusts the rate at which calcium and magnesium ions are lost in the urine. PTH also increases the movement of phosphate ions from the blood to urine for excretion.

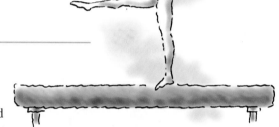

The main function of PTH is to help regulate the blood's calcium balance.

Adrenal glands

The two *adrenal glands* each lie on top of a kidney. These almond-shaped glands contain two distinct structures—the adrenal cortex and the adrenal medulla—that function as separate endocrine glands.

Adrenal cortex

The *adrenal cortex* is the large outer layer. It forms the bulk of the adrenal gland. It has three zones, or cell layers:

1. The *zona glomerulosa,* the outermost zone, produces mineralocorticoids (primarily aldosterone) that help maintain fluid balance by increasing sodium reabsorption.
2. The *zona fasciculata,* the middle and largest zone, produces the glucocorticoids cortisol (hydrocortisone), cortisone, and corticosterone as well as small amounts of the sex hormones androgen and estrogen. Glucocorticoids help regulate metabolism and resistance to stress.
3. The *zona reticularis,* the innermost zone, produces some sex hormones (DHEA and DHEA sulfate); this occurs at the time of "adrenarche."

Adrenal medulla

The *adrenal medulla,* or inner layer of the adrenal gland, functions as part of the sympathetic nervous system and produces two catecholamines: epinephrine and norepinephrine. Because catecholamines play an important role in the autonomic nervous system (ANS), the adrenal medulla is considered a neuroendocrine structure.

Memory jogger

To remember the location of the adrenal glands, think **Add-Renal**. They're "added" to the renal organs, the kidneys.

Pancreas

The *pancreas,* a triangular organ, is nestled in the curve of the duodenum, stretching horizontally behind the stomach and extending to the spleen.

Endo and exo

The pancreas performs both endocrine and exocrine functions. As its endocrine function, the pancreas secretes hormones, while its exocrine function is secreting digestive enzymes. *Acinar cells* make up most of the gland and regulate pancreatic exocrine function. (See *Pancreas stimulation.*)

Now I get it!

Pancreas stimulation

Many endocrine cells possess receptors on their membranes that respond to stimuli:
• Neuron stimulation of pancreatic beta cells (shown below) causes production of the hormone precursor preproinsulin.
• Preproinsulin is converted to proinsulin in beadlike ribosomes located on the endoplasmic reticulum.

• Proinsulin is transferred to the Golgi complex, which collects it into secretory granules and converts it to insulin.
• Secretory granules fuse with the plasma membrane and disperse insulin into the bloodstream.
• Hormonal release by membrane fusion is called *exocytosis.*

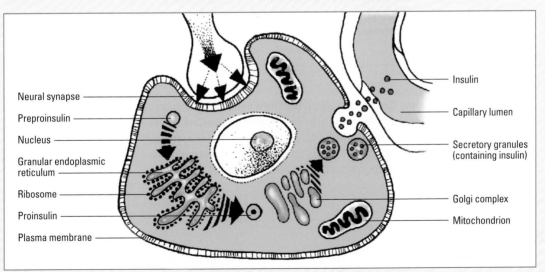

Islands in an acinar sea

The endocrine cells of the pancreas are called the *islet cells,* or *islets of Langerhans*. These cells exist in clusters and are found scattered among the acinar cells. The islets contain alpha, beta, and delta cells that produce important hormones:

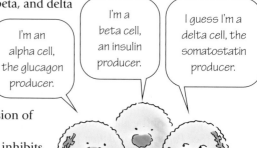

- Alpha cells produce *glucagon,* a hormone that raises the blood glucose level by triggering the breakdown of glycogen to glucose.
- Beta cells produce *insulin*. Insulin lowers the blood glucose level by stimulating the conversion of glucose to glycogen.
- Delta cells produce *somatostatin*. Somatostatin inhibits the release of GH, corticotropin, and certain other hormones.

Thymus

The *thymus* is located below the sternum and contains lymphatic tissue. It reaches maximal size at puberty and then starts to atrophy.

Producing Mr. T cells, fool!

Because the thymus produces T cells, which are important in cell-mediated immunity, its major role seems to be related to the immune system. However, the thymus also produces the peptide hormones thymosin and thymopoietin. These hormones promote growth of peripheral lymphoid tissue.

Pineal gland

The tiny *pineal gland* lies at the back of the third ventricle of the brain. It produces the hormone *melatonin,* primarily during the dark hours of the day. Little hormone is produced during daytime hours, thereby influencing sleep-wake cycles. Melatonin is thought to regulate circadian rhythms, body temperature, cardiovascular function, and reproduction.

Gonads

The *gonads* include the ovaries (in females) and the testes (in males).

Ovaries

The *ovaries* are paired, oval glands that are situated on either side of the uterus. They produce ova (eggs) and the steroidal hormones estrogen and progesterone. These hormones have four functions:

1. They promote development and maintenance of female sex characteristics.
2. They regulate the menstrual cycle.
3. They maintain the uterus for pregnancy.
4. Along with other hormones, they prepare the mammary glands for lactation.

Testes

The *testes* are paired structures that lie in an extra-abdominal pouch (scrotum) in the male. They produce spermatozoa and the male sex hormone testosterone. Testosterone stimulates and maintains masculine sex characteristics and triggers the male sex drive.

Hormones

Hormones are complex chemical substances that trigger or regulate the activity of an organ or a group of cells. Hormones are classified by their molecular structure as polypeptides, steroids, or amines.

Polypeptides

Polypeptides are protein compounds made of many amino acids that are connected by peptide bonds. They include:

- anterior pituitary hormones (GH, TSH, corticotropin, FSH, LH, and prolactin)
- posterior pituitary hormones (ADH and oxytocin)
- parathyroid hormone (PTH)
- pancreatic hormones (insulin and glucagon).

Steroids

Steroids are derived from cholesterol. They include:

- adrenocortical hormones secreted by the adrenal cortex (aldosterone and cortisol)
- sex hormones secreted by the gonads (estrogen and progesterone in females and testosterone in males).

Amines

Amines are derived from *tyrosine,* an essential amino acid found in most proteins. They include:
- thyroid hormones (T_4 and T_3)
- catecholamines (epinephrine, norepinephrine, and dopamine).

Hormone release and transport

Although all hormone release results from endocrine gland stimulation, release patterns of hormones vary greatly. For example:
- Corticotropin (secreted by the anterior pituitary) and cortisol (secreted by the adrenal cortex) are released in spurts in response to body rhythm cycles. Levels of these hormones peak in the morning.
- Secretion of PTH (by the parathyroid gland) and prolactin (by the anterior pituitary) occurs fairly evenly throughout the day.
- Secretion of insulin by the pancreas can occur at a steady rate or sporadically, depending on blood glucose levels.

Hormonal action

When a hormone reaches its target site, it binds to a specific receptor on the cell membrane or within the cell. Polypeptides and some amines bind to membrane receptor sites. The smaller, more lipid-soluble steroids and thyroid hormones diffuse through the cell membrane and bind to intracellular receptors.

Right on target!

After binding occurs, each hormone produces unique physiologic changes, depending on its target site and its specific action at that site. A particular hormone may have different effects at different target sites.

Hormonal regulation

To maintain the body's delicate equilibrium, a feedback mechanism regulates hormone production and secretion. The mechanism involves hormones, blood chemicals and metabolites, and the nervous system. This system may be simple or complex. (See *The feedback loop.*)

Thyroid and steroid hormones circulate while bound to plasma proteins, whereas catecholamines and most polypeptides aren't protein bound.

Now I get it!

The feedback loop

This diagram shows the negative feedback mechanism that helps regulate the endocrine system.

From simple...

Simple feedback occurs when the level of one substance regulates the secretion of hormones (simple loop). For example, a low serum calcium level stimulates the parathyroid gland to release parathyroid hormone (PTH). PTH, in turn, promotes resorption of calcium from the GI tract, kidneys, and bones. A high serum calcium level inhibits PTH secretion.

...to complex

Hypothalamic stimulation can also trigger a complex feedback mechanism. First, the hypothalamus sends releasing and inhibiting factors or hormones to the anterior pituitary. In response, the anterior pituitary secretes tropic hormones, such as growth hormone (GH), prolactin (PRL), corticotropin, thyroid-stimulating hormone (TSH), follicle-stimulating hormone (FSH), and luteinizing hormone (LH). At the appropriate target gland, these hormones stimulate the target organ to release other hormones that regulate various body functions. When these hormones reach normal levels in body tissue, a feedback mechanism inhibits further hypothalamic and pituitary secretion.

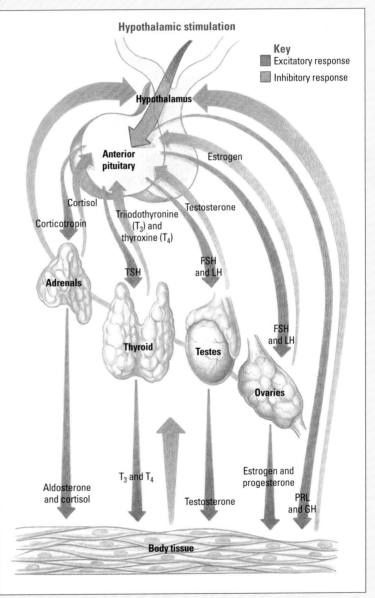

Signaling secretory cells

For normal function, each gland must contain enough appropriately programmed secretory cells to release active hormones on demand. Secretory cells need supervision. A secretory cell can't sense on its own when to release the hormone or how much to release. It gets this information from sensing and signaling systems that integrate many messages. Together, stimulatory and inhibitory signals actively control the rate and duration of hormone release.

When released, the hormone travels to *target cells,* where a receptor molecule recognizes it and binds to it. (See *A close look at target cells.*)

Now I get it!

A close look at target cells

A hormone acts only on cells that have receptors specific to that hormone. The sensitivity of a target cell depends on how many receptors it has for a particular hormone. The more receptor sites, the more sensitive the target cell.

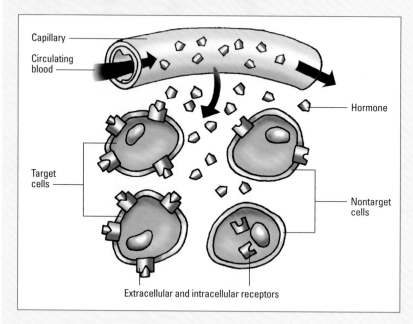

Capillary

Circulating blood

Hormone

Target cells

Nontarget cells

Extracellular and intracellular receptors

A hormone acts only on a cell that has a receptor specific to that hormone.

Mechanisms that control hormone release

Four basic mechanisms control hormone release:
1. the pituitary-target gland axis
2. the hypothalamic-pituitary-target gland axis
3. chemical regulation
4. nervous system regulation.

Pituitary-target gland axis

The pituitary gland regulates other endocrine glands—and their hormones—through secretion of *trophic hormones* (releasing and inhibiting hormones). These hormones include:
- corticotropin, which regulates adrenocortical hormones
- TSH, which regulates T_4 and T_3
- LH, which regulates gonadal hormones.

Picking up feedback

The pituitary gland gets feedback about target glands by continuously monitoring levels of hormones produced by these glands. If a change occurs, the pituitary gland corrects it in one of two ways:
- by increasing the trophic hormones, which stimulate the target gland to increase production of target gland hormones
- by decreasing the trophic hormones, thereby decreasing target gland stimulation and target gland hormone levels.

Hypothalamic-pituitary-target gland axis

The hypothalamus also produces trophic hormones that regulate anterior pituitary hormones. By controlling anterior pituitary hormones, which regulate the target gland hormones, the hypothalamus affects target glands as well.

Chemical regulation

Endocrine glands not controlled by the pituitary gland may be controlled by specific substances that trigger gland secretions. For example, blood glucose level is a major regulator of glucagon and insulin release. When blood glucose level rises, the pancreas is stimulated to increase insulin secretion and suppress glucagon secretion. A depressed level of blood glucose, on the other hand, triggers increased glucagon secretion and suppresses insulin secretion. (See *Endocrine changes with aging,* page 122.)

It says here that specific substances, such as blood glucose, can trigger gland secretions.

Endocrine changes with aging

As a person ages, normal changes in endocrine function takes place. Sex hormone levels decrease (testosterone, progesterone, and estrogen levels). There is a decline in serum aldosterone levels causing orthostatic hypotension, and PTH rises, which can contribute to osteoporosis. There can also be a rise in FSH, LH, and norepinephrine levels.

Another common and important endocrine change in elderly people is a change in glucose metabolism in response to stress. Normally, both young adults and elderly people have similar fasting blood glucose levels. However, under stressful conditions, an elderly person's blood glucose level rises higher and remains elevated longer than does a younger adult's.

Nervous system regulation

The central nervous system (CNS) helps to regulate hormone secretion in several ways.

Hypothalamus has control...

The hypothalamus controls pituitary hormones. Because hypothalamic nerve cells stimulate the posterior pituitary to secrete ADH and oxytocin, these hormones are controlled directly by the CNS.

...but stimuli matter, too

Nervous system stimuli—such as hypoxia (oxygen deficiency), nausea, pain, stress, and certain drugs—also affect ADH levels.

ANS steers this ship...

The ANS controls catecholamine secretion by the adrenal medulla.

...while stress spikes corticotropin

The nervous system also affects other endocrine hormones. For example, stress, which leads to sympathetic stimulation, causes the pituitary to release corticotropin.

Quick quiz

1. The purpose of the endocrine system is to:
    A. deliver nutrients to the body's cells.
    B. regulate and integrate the body's metabolic activities.
    C. eliminate waste products from the body.
    D. control the body's temperature and produce blood cells.

    Answer: B. Along with the nervous system, the endocrine system regulates and integrates the body's metabolic activities.

2. The mechanism that helps regulate the endocrine system is called the:
    A. transport mechanism.
    B. self-regulation mechanism.
    C. feedback mechanism.
    D. pituitary-target gland axis.

    Answer: C. The negative feedback mechanism helps regulate the endocrine system by signaling to the endocrine glands the need for changes in hormone levels.

3. The gland that produces glucagon is the:
    A. pancreas.
    B. thymus.
    C. adrenal gland.
    D. pituitary gland.

    Answer: A. The alpha cells of the pancreas produce glucagon, a hormone that raises the blood glucose level by triggering the breakdown of glycogen to glucose.

4. Pituitary hormones are controlled by the:
    A. pancreas.
    B. hypothalamus.
    C. thyroid gland.
    D. parathyroid glands.

    Answer: B. The hypothalamus controls pituitary hormones.

Scoring

☆☆☆   If you answered all four questions correctly, astonishing! Your brain cells must be on steroids!

  ☆☆   If you answered three questions correctly, bon voyage! You've just won a trip to the islets of Langerhans!

    ☆   If you answered fewer than three questions correctly, don't moan over these hormones. Focus your energy on the chapter ahead.

Just for fun!

The two adrenal glands each lie on top of a kidney. These almond-shaped glands contain two distinct structures—the adrenal cortex and the adrenal medulla—that function as separate endocrine glands. Unscramble the words on the left to reveal the three zones, or cells layers, of the adrenal cortex. Then draw a line from each box to the specific characteristics of each zone, listed on the right.

AZON OAGLESOLRUM

_ _ _ _ _ _ _
_ _ _ _ _ _ _ _ _ _ _

NAZO ACATUSFAIL

_ _ _ _ _ _
_ _ _ _ _ _ _ _ _ _ _

NOZA CURLERSIAIT

_ _ _ _ _ _
_ _ _ _ _ _ _ _ _ _ _

A. Outermost zone
B. Helps regulate metabolism and resistance to stress
C. Produces mainly glucocorticoids and some sex hormones
D. Helps maintain fluid balance by increasing sodium reabsorption
E. Produces mineralocorticoids (primarily aldosterone)
F. Produces hydrocortisone, cortisone, and corticosterone
G. Middle and largest zone
H. Innermost zone
I. Produces small amounts of sex hormones androgen and estrogen

Answer: ZONA GLOMERULOSA: A, D, E; ZONA FASCICULATA: B, F, G, I; ZONA RETICULARIS: C, H

Selected References

Aging Changes in Hormone Production. (2014). *Medline Plus*. Retrieved from https://www.nlm.nih.gov/medlineplus/ency/article/004000.htm

Berg, J. M., Tymoczko, J. L., & Stryer, L. (2002). *Biochemistry* (5th ed.). Retrieved from http://www.ncbi.nlm.nih.gov/books/NBK22339/

Bird, I. (2012). In the zone: Understanding zona reticularis function and its transformation by adrenarche. *Journal of Endocrinology, 214*, 109–111. doi: 10.1530/JOE-12-0246

Hormone. (2016). *Encyclopædia Britannica*. Retrieved from https://www.britannica.com/science/hormone

How Does The Thyroid Work? (2015). *PubMed Health*. Retrieved from http://www.ncbi.nlm.nih.gov/pubmedhealth/PMH0072572/

Huether, S. E., & McCance, K. L. (2008). *Understanding pathophysiology*. St. Louis, MO: Mosby Elsevier.

Pituitary Gland Introduction. (2016). *Emory Healthcare*. Retrieved from http://www.emoryhealthcare.org/pituitary/pituitary-gland.html

The Pancreas Has Two Functional Components. (2015). *Johns Hopkins Medicine*. Retrieved from http://pathology.jhu.edu/pancreas/BasicOverview3.php?area=ba

Thymus. (2016). *Endocrine Society*. Retrieved from http://www.endocrine.org/news-room/glossary/thymus-to-z

Chapter 8

Cardiovascular system

Just the facts

In this chapter, you'll learn:

◆ structures of the heart and their functions
◆ the heart's conduction system
◆ the flow of blood through the heart and the body.

A look at the cardiovascular system

The cardiovascular system (sometimes called the *circulatory system*) consists of the *heart, blood vessels,* and *lymphatics.* This network brings life-sustaining oxygen and nutrients to the body's cells, removes metabolic waste products, and carries hormones from one part of the body to another.

Doing double duty

The heart is actually two separate pumps: the right side pumps the blood to the lungs to receive oxygen, and the left side pumps the oxygenated blood to the rest of the body.

Where the heart lies

About the size of a closed fist and weighing less than a pound, the heart lies beneath the sternum in the *mediastinum* (the cavity between the lungs), between the second and sixth ribs. In most people, the heart rests obliquely, with its right side below and almost in front of the left. Because of its oblique angle, the heart's broad part or top is at its upper right, and its pointed end (apex) is at its lower left. The apex is the *point of maximal impulse,* where the heart sounds are the loudest.

The heart pumps around 2,000 gallons of blood throughout the body each day, beating approximately 100,000 times!

Heart structure

Surrounded by a protective sac called the *pericardium,* the heart has a wall made up of three layers: the *myocardium, endocardium,* and *epicardium.* Within the heart lie four chambers (two atria and two ventricles) and four valves (two atrioventricular [AV] and two semilunar valves). (See *Inside the heart.*)

I'm supported and protected by a tough, fibrous sac, but I stay comfortable because it has a smooth inner lining.

Pericardium

The *pericardium* is a fibroserous sac that surrounds the heart and the roots of the *great vessels* (those vessels that enter and leave the heart). It consists of the fibrous pericardium and the serous pericardium.

Fibrous fits freely

The *fibrous pericardium,* composed of tough, white fibrous tissue, fits loosely around the heart, protecting it.

Serous is smooth

The *serous pericardium,* the thin, smooth inner portion, has two layers:
- The *parietal layer* lines the inside of the fibrous pericardium.
- The *visceral layer* adheres to the surface of the heart.

The space between

Between the fibrous and serous pericardium is the *pericardial cavity.* This space contains *pericardial fluid* that lubricates the surfaces of the space and allows the heart to move easily during contraction.

The wall

The wall of the heart consists of three layers:
1. The *epicardium,* the outer layer (and the visceral layer of the serous pericardium), is made up of squamous epithelial cells overlying connective tissue.
2. The *myocardium,* the middle (and thickest) layer, forms most of the heart wall. It has striated muscle fibers that cause the heart to contract.

Inside the heart

Within the heart lie four chambers (two atria and two ventricles) and four valves (two atrioventricular and two semilunar valves). A system of blood vessels carries blood to and from the heart.

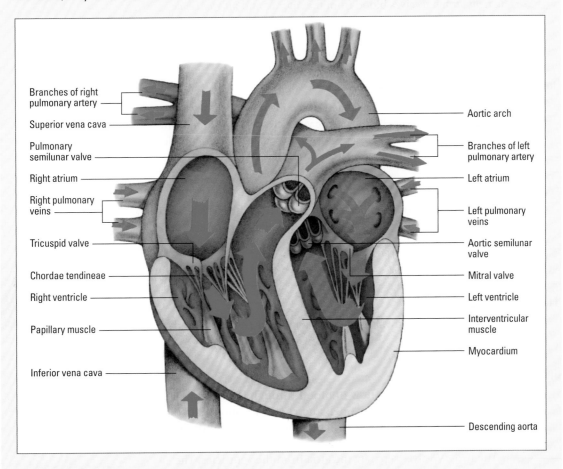

Branches of right pulmonary artery

Superior vena cava

Pulmonary semilunar valve

Right atrium

Right pulmonary veins

Tricuspid valve

Chordae tendineae

Right ventricle

Papillary muscle

Inferior vena cava

Aortic arch

Branches of left pulmonary artery

Left atrium

Left pulmonary veins

Aortic semilunar valve

Mitral valve

Left ventricle

Interventricular muscle

Myocardium

Descending aorta

3. The *endocardium,* the heart's inner layer, consists of endothelial tissue with small blood vessels and bundles of smooth muscle.

The chambers

The heart contains four hollow chambers: two atria (singular: atrium) and two ventricles.

Upstairs...

The *atria*, the upper chambers, are separated by the *interatrial septum*. They receive blood returning to the heart and supply blood to the ventricles.

...where the blood comes in

The *right atrium* receives blood from the *superior* and *inferior venae cavae*. The *left atrium*, which is smaller but has thicker walls than does the right atrium, forms the uppermost part of the heart's left border. It receives blood from the four pulmonary veins.

Downstairs...

The *right* and *left ventricles*, separated by the *interventricular septum*, make up the two lower chambers. The ventricles receive blood from the atria. Composed of highly developed musculature, the ventricles are larger and have thicker walls than do the atria.

...where the blood goes out

The right ventricle pumps blood to the lungs. The left ventricle, which is larger than the right, pumps blood through all other vessels of the body.

The valves

The heart contains four valves, two *atrioventricular (AV) valves* and two *semilunar valves*.

One way only

The valves allow forward flow of blood through the heart and prevent backward flow. They open and close in response to pressure changes caused by ventricular contraction and blood ejection.

The two AV valves separate the atria from the ventricles. The right AV valve, called the *tricuspid valve*, prevents backflow from the right ventricle into the right atrium. The left AV valve, called the *mitral valve*, prevents backflow from the left ventricle into the left atrium.

One of the two semilunar valves is the *pulmonic valve*, which prevents backflow from the pulmonary artery into the right ventricle. The other semilunar valve is the *aortic valve*, which prevents backflow from the aorta into the left ventricle.

Memory jogger

If you can remember that there are two distinct heart sounds, you can recall that there are two sets of heart valves. Closure of the atrioventricular valves makes the first heart sound, the **lub**; closure of the semilunar valves makes the second heart sound, the **dub**.

On the cusps

The tricuspid valve has three triangular *cusps,* or leaflets. The mitral valve, also called the *bicuspid valve,* contains two cusps, a large anterior and a smaller posterior. *Chordae tendineae* attach the cusps of the AV valves to papillary muscles in the ventricles. The semilunar valves have three cusps that are shaped like half-moons.

Conduction system

Contraction of the heart, occurring as a result of its *conduction system,* causes blood to move throughout the body. (See *Cardiac conduction system.*)

Setting the pace

The conduction system of the heart contains *pacemaker cells,* which have three unique characteristics:
- *automaticity,* the ability to generate an electrical impulse spontaneously

Cardiac conduction system

Specialized fibers propagate electrical impulses throughout the heart's cells, causing the muscle to contract. This illustration shows the elements of the cardiac conduction system.

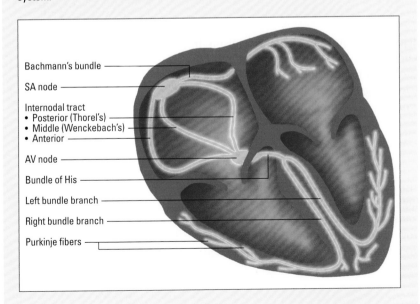

Bachmann's bundle

SA node

Internodal tract
- Posterior (Thorel's)
- Middle (Wenckebach's)
- Anterior

AV node

Bundle of His

Left bundle branch

Right bundle branch

Purkinje fibers

Electrical impulses help me conduct myself at a special rhythm.

- *excitability,* the ability to respond to an impulse
- *conductivity,* the ability to pass the impulse to the next cell.

Feeling impulsive

The *sinoatrial (SA) node,* located on the *endocardial surface* of the right atrium, near the superior vena cava, is the normal pacemaker of the heart, generating an impulse between 60 and 100 times per minute. The firing of the SA node spreads an impulse throughout the right and left atria, resulting in atrial contraction.

Fill 'er up

The *AV node,* situated low in the *septal wall* of the right atrium, slows impulse conduction between the atria and ventricles. This "resistor" node allows time for the contracting atria to fill the ventricles with blood before the lower chambers contract.

My conduction system has two backup impulse generators.

Spreading the word—"contract"

From the AV node, the impulse travels to the **bundle of His** (modified muscle fibers), branching off to the right and left bundles. Finally, the impulse travels to the *Purkinje fibers,* the distal portions of the left and right bundle branches. These fibers fan across the surface of the ventricles from the endocardium to the myocardium. As the impulse spreads, it brings "the word" to the blood-filled ventricles to contract.

Foolproof

The conduction system has two built-in safety mechanisms. If the SA node fails to fire, the AV node will generate an impulse between 40 and 60 times per minute. If both the SA node and AV node fail, the ventricles can generate their own impulse between 20 and 40 times per minute.

Cardiac cycle

The *cardiac cycle* is the period from the beginning of one heartbeat to the beginning of the next. During this cycle, electrical and mechanical events must occur in the proper sequence and to a precise degree to provide adequate cardiac output to the body. The cardiac cycle has two phases: *systole* and *diastole.* (See *Events in the cardiac cycle.*)

Now I get it!

Events in the cardiac cycle

The cardiac cycle consists of the following five events.

1. Isovolumetric ventricular contraction—In response to ventricular depolarization, tension in the ventricles increases. This rise in pressure within the ventricles leads to closure of the mitral and tricuspid valves. The pulmonic and aortic valves stay closed during the entire phase.

2. Atrial systole—Known as the atrial kick, atrial systole (coinciding with late ventricular diastole) supplies the ventricles with the remaining 30% of the blood for each heartbeat.

3. Ventricular filling—Atrial pressure exceeds ventricular pressure, which causes the mitral and tricuspid valves to open. Blood then flows passively into the ventricles. About 70% of ventricular filling takes place during this phase.

4. Isovolumetric relaxation—When ventricular pressure falls below the pressure in the aorta and pulmonary artery, the aortic and pulmonic valves close. All valves are closed during this phase. Atrial diastole occurs as blood fills the atria.

5. Ventricular ejection—When ventricular pressure exceeds aortic and pulmonary arterial pressure, the aortic and pulmonic valves open and the ventricles eject blood.

Contract...

At the beginning of *systole,* the ventricles contract. Increasing blood pressure in the ventricles forces the AV valves (mitral and tricuspid) to close and the semilunar valves (pulmonic and aortic) to open.

As the ventricles contract, ventricular blood pressure builds until it exceeds the pressure in the pulmonary artery and the aorta. Then the semilunar valves open, and the ventricles eject blood into the aorta and the pulmonary artery.

...and release

When the ventricles empty and relax, ventricular pressure falls below the pressure in the pulmonary artery and the aorta. At the beginning of *diastole*, the semilunar valves close to prevent the backflow of blood into the ventricles, and the mitral and tricuspid valves open, allowing blood to flow into the ventricles from the atria.

When the ventricles become full near the end of this phase, the atria contract to send the remaining blood to the ventricles. A new cardiac cycle begins as the heart enters systole again.

Normal cardiac output is around 5 L per minute at rest.

Cardiac output

Cardiac output refers to the amount of blood the heart pumps in 1 minute. It's equal to the heart rate multiplied by the *stroke volume*, the amount of blood ejected with each heartbeat. Stroke volume, in turn, depends on three major factors: *preload, contractility,* and *afterload*. (See *Understanding preload, contractility, and afterload*.)

Blood flow

As blood makes its way through the vascular system, it travels through five distinct types of blood vessels, involving three methods of circulation.

Blood vessels

The five types of blood vessels are arteries, arterioles, capillaries, venules, and veins. The structure of each type of vessel differs according to its function in the cardiovascular system and the pressure exerted by the volume of blood at various sites within the system.

Through thick...

Arteries have thick, muscular walls to accommodate the flow of blood at high speeds and pressures. *Arterioles* have thinner walls than do arteries. They constrict or dilate to control blood flow to the *capillaries*, which (being microscopic) have walls composed of only a single layer of endothelial cells

Understanding preload, contractility, and afterload

If you think of the heart as a balloon, it will help you understand stroke volume.

Blowing up the balloon

Preload is the stretching of muscle fibers in the ventricles. This stretching results from blood volume in the ventricles at end-diastole. According to *Starling's law,* the more the heart muscles stretch during diastole, the more forcefully they contract during systole. Think of preload as the balloon stretching as air is blown into it. The more the air, the greater the stretch.

The balloon's stretch

Contractility refers to the inherent ability of the myocardium to contract normally. Contractility is influenced by preload. The more the air in the balloon, the greater the stretch, and the farther the balloon will fly when air is allowed to expel.

The knot that ties the balloon

Afterload refers to the pressure that the ventricular muscles must generate to overcome the higher pressure in the aorta to get the blood out of the heart. Think of it as the knot on the end of the balloon, which it has to work against to get the air out.

...and thin

Venules gather blood from the capillaries; their walls are thinner than those of arterioles. *Veins* have thinner walls than do arteries but have larger diameters because of the low blood pressures of venous return to the heart.

Taking the long way home

About 60,000 miles of arteries, arterioles, capillaries, venules, and veins keep blood circulating to and from every functioning cell in the body. (See *Major blood vessels,* page 134.)

Body shop

Major blood vessels

This illustration shows the body's major arteries and veins.

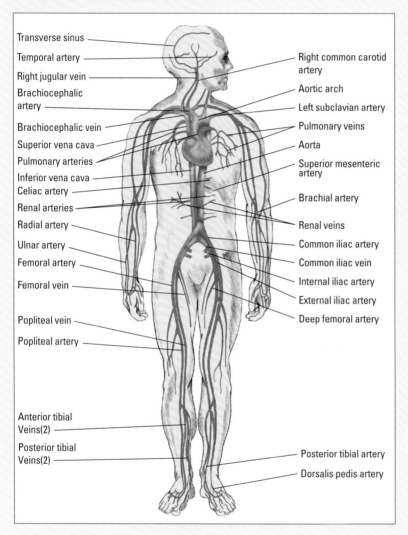

Transverse sinus

Temporal artery

Right jugular vein

Brachiocephalic artery

Brachiocephalic vein

Superior vena cava

Pulmonary arteries

Inferior vena cava

Celiac artery

Renal arteries

Radial artery

Ulnar artery

Femoral artery

Femoral vein

Popliteal vein

Popliteal artery

Anterior tibial Veins(2)

Posterior tibial Veins(2)

Right common carotid artery

Aortic arch

Left subclavian artery

Pulmonary veins

Aorta

Superior mesenteric artery

Brachial artery

Renal veins

Common iliac artery

Common iliac vein

Internal iliac artery

External iliac artery

Deep femoral artery

Posterior tibial artery

Dorsalis pedis artery

The human body has nearly 60,000 miles of blood vessels. That's enough to circle the globe 2½ times!

Circulation

There are three methods of circulation that carry blood throughout the body: *pulmonary, systemic,* and *coronary.*

Pulmonary circulation

Blood travels to the lungs to pick up oxygen and release carbon dioxide.

Returns and exchanges

As the blood moves from the heart, to the lungs, and back again, it proceeds as follows:

- Unoxygenated blood travels from the right ventricle through the pulmonic valve into the *pulmonary arteries.*
- Blood passes through progressively smaller arteries and arterioles into the capillaries of the lungs.
- Blood reaches the *alveoli* and exchanges carbon dioxide for oxygen.
- Oxygenated blood then returns via venules and veins to the *pulmonary veins,* which carry it back to the heart's left atrium.

At rest, only about 20% of my blood goes to skeletal muscles. When I exercise, that percentage can increase to 70%.

Systemic circulation

Blood pumped from the left ventricle carries oxygen and other nutrients to body cells and transports waste products for excretion.

Branching out

The major artery, the *aorta,* branches into vessels that supply specific organs and areas of the body. As it arches out of the top of the heart and down to the abdomen, three arteries branch off the top of the arch to supply the upper body with blood:

- the *left common carotid artery*
- the *left subclavian artery*
- the *brachiocephalic artery, which* further divides into the right common carotid and right subclavian arteries.

As the aorta descends through the thorax and abdomen, its branches supply the organs of the GI and genitourinary systems, spinal column, and lower chest and abdominal muscles. Then the aorta divides into the *iliac arteries,* which further divide into *femoral arteries.*

Division = addition = perfusion

As the arteries divide into smaller units, the number of vessels increases dramatically, thereby increasing the area of tissue to which blood flows, also called the *area of perfusion.*

Dilation is another part of the equation

At the end of the arterioles and the beginning of the capillaries, strong *sphincters* control blood flow into the tissues. These sphincters dilate to permit more flow when needed, close to shunt blood to other areas, or constrict to increase blood pressure.

A large area of low pressure

Although the *capillary bed* contains the smallest vessels, it supplies blood to the largest number of cells. Capillary pressure is extremely low to allow for the exchange of nutrients, oxygen, and carbon dioxide with body cells. From the capillaries, blood flows into venules and, eventually, into veins.

No backflow

Valves in the veins prevent blood backflow. Pooled blood in each valved segment is moved toward the heart by pressure from the moving volume of blood from below. The veins merge until they form two main branches, the *superior vena cava* and *inferior vena cava*, that return blood to the right atrium.

Coronary circulation

The heart relies on the coronary arteries and their branches for its supply of oxygenated blood and depends on the cardiac veins to remove oxygen-depleted blood. (See *Vessels that supply the heart.*)

The heart gets its part

During systole, blood is ejected into the aorta from the left ventricle. During diastole, blood flows out of the heart and then through the coronary arteries to nourish the heart muscle.

From the right...

The *right coronary artery* supplies blood to the right atrium, part of the left atrium, most of the right ventricle, and the inferior part of the left ventricle. It also supplies the blood to both the SA and AV nodes.

...and from the left

The *left coronary artery,* which splits into the left *anterior descending artery* and *circumflex artery,* supplies blood to the left atrium, most of the left ventricle, and most of the interventricular septum.

Like any other muscle, I also need oxygen to function. Ah, there's nothing like a little fresh air!

Zoom in

Vessels that supply the heart

Coronary circulation involves the arterial system of blood vessels that supply oxygenated blood to the heart and the venous system that removes oxygen-depleted blood from it.

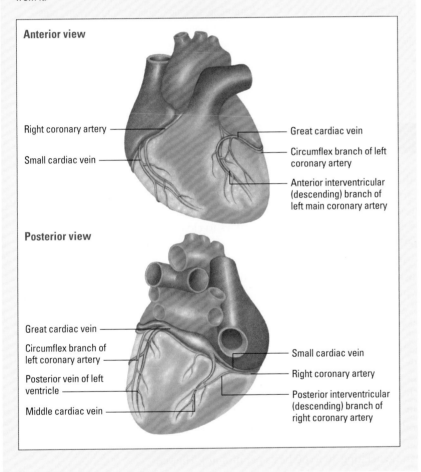

Anterior view

Right coronary artery

Small cardiac vein

Great cardiac vein

Circumflex branch of left coronary artery

Anterior interventricular (descending) branch of left main coronary artery

Posterior view

Great cardiac vein

Circumflex branch of left coronary artery

Posterior vein of left ventricle

Middle cardiac vein

Small cardiac vein

Right coronary artery

Posterior interventricular (descending) branch of right coronary artery

As a coronary artery, my job is to supply the heart with the oxygen it needs to keep beating.

Superficially speaking

The *cardiac veins* lie superficial to the arteries. The largest vein, the *coronary sinus,* opens into the right atrium. Most of the major cardiac veins empty into the coronary sinus, except for the *anterior cardiac veins,* which empty into the right atrium.

Senior moment

Cardiovascular changes with aging

Due to loss of cardiac cells with age, the pacemaker of the heart may not be as efficient as it used to be, which may result in a lower heart rate.

Arteries become more rigid, causing systolic blood pressure to rise.

The left ventricular wall may grow thicker from an increased effort to pump blood.

Heart valves also become thicker from fibrotic and sclerotic changes, which can cause narrowing and/or prevent the valves from closing completely.

Baroreceptors help regulate blood flow and pressure with movement. This function diminishes with age, as well, and can lead to low pulse rate, dizziness, and fainting.

Quick quiz

1. During systole, the ventricles contract. This causes:
   A. all four heart valves to close.
   B. the AV valves to close and the semilunar valves to open.
   C. the AV valves to open and the semilunar valves to close.
   D. all four heart valves to open.

Answer: B. During systole, the pressure is greater in the ventricles than in the atria, causing the AV valves (the tricuspid and mitral valves) to close. The pressure in the ventricles is also greater than the pressure in the aorta and pulmonary artery, forcing the semilunar valves (the pulmonic and aortic valves) to open.

2. The normal pacemaker of the heart is:
   A. the SA node.
   B. the AV node.
   C. the ventricles.
   D. the Purkinje fibers.

Answer: A. The SA node is the normal pacemaker of the heart, generating impulses 60 to 100 times per minute. The AV node is the secondary pacemaker of the heart (generating 40 to 60 beats per minute). The ventricles are the last line of defense (generating 20 to 40 beats per minute).

3. The pressure the ventricular muscle must generate to overcome the higher pressure in the aorta refers to:
   A. contractility.
   B. preload.
   C. blood pressure.
   D. afterload.

Answer: D. Afterload is the pressure the ventricular muscle must generate to overcome the higher pressure in the aorta to get the blood out of the heart.

4.   The vessels that carry oxygenated blood back to the heart and left
atrium are:
   A.   capillaries.
   B.   pulmonary veins.
   C.   pulmonary arteries.
   D.   superior and inferior venae cavae.

Answer: B. Oxygenated blood returns by way of venules and veins to
the pulmonary veins, which carry it back to the heart's left atrium.

5.   The layer of the heart responsible for contraction is the:
   A.   myocardium.
   B.   pericardium.
   C.   endocardium.
   D.   epicardium.

Answer: A. The myocardium has striated muscle fibers that cause the
heart to contract.

Just for fun!

Stroke volume depends on three major factors. Unscramble the jumbled words on the left to reveal the
names of these three factors. Then draw lines from each of these boxes to the descriptions on the right
that match each term.

PADROLE

_ _ _ _ _ _ _ _ _ _ _ _ _ _ _ _

A. The balloon's stretch
B. Pressure ventricular muscles must generate to
   overcome aortic pressure
C. Like blowing up a balloon
D. The balloon's knot
E. Inherent ability of myocardium to contract normally
F. Stretching of muscle fiber in ventricles

PACRYLICTINTTO

_ _ _ _ _ _ _ _ _ _ _ _ _ _ _ _

AFLOATRED

_ _ _ _ _ _ _ _ _ _ _ _ _ _ _ _

Answer: PRELOAD: C, F; CONTRACTILITY: A, E; AFTERLOAD: B, D

Scoring

☆☆☆ If you answered all five questions correctly, marvelous! You've gotten to the heart of the cardiovascular system.

☆☆ If you answered four questions correctly, great! We won't call you "vein" if you're a little proud of yourself.

☆ If you answered fewer than four questions correctly, take heart! It's time to circulate on to the next chapter.

Selected References

Cedars-Sinai. (2016). *Anatomy of the aorta and heart.* Retrieved from http://cedars-sinai.edu/Patients/Programs-and-Services/Heart-Institute/Centers-and-Programs/Aortic- Program/Anatomy-of-the-Aorta-and-Heart.aspx

The Franklin Institute. (2016). *Blood vessels.* Retrieved from https://www.fi.edu/heart/blood-vessels

The John Hopkins Health System. (n.d.). *Anatomy and function of the coronary arteries.* Retrieved from http://www.hopkinsmedicine.org/healthlibrary/conditions/cardiovascular_diseases/anatomy_and_function_of_the_coronary_arteries_85,p00196/

Mahadevan, V. (2015). Anatomy of the heart. *Surgery, 33*(2), 47–51. doi: 10.1016/j.mpsur.2014.12.001.

McCance, K., & Huether, S. (2010). *Pathophysiology: The biologic basis for disease in adults and children* (6th ed.). Maryland Heights, MO: Mosby.

National Heart, Lung, and Blood Institute. (2011). *What is the heart?* Retrieved from https://www.nhlbi.nih.gov/health/health-topics/topics/hhw

National Institute on Aging. (n.d.). *Blood vessels and aging: The rest of the journey.* Retrieved from https://www.nia.nih.gov/health/publication/aging-hearts-and-arteries/chapter-4- blood- vessels-and-aging-rest-journey

Texas Heart Institute. (2015). *Heart anatomy.* Retrieved from http://www.texasheart.org/HIC/Anatomy/anatomy2.cfm

U.S. National Library of Medicine. (2016). *Aging changes in the heart and blood vessels.* Retrieved from https://www.nlm.nih.gov/medlineplus/ency/article/004006.htm

Vincent, J.-L. (2008). Understanding cardiac output. *Critical Care, 12*(4), 174. doi: 10.1186/cc6975.

Whitaker, R. H. (2014). Anatomy of the heart. *Medicine, 42*(8), 406–408. doi: 10.1016/j.mpmed.2014.05.007.

Chapter 9

Hematologic system

Just the facts

In this chapter, you'll learn:

♦ the way in which blood cells develop

♦ functions of the different blood components

♦ the way in which blood cells clot

♦ blood groups and their significance.

A look at the hematologic system

The hematologic system consists of the blood and bone marrow. Blood delivers oxygen and nutrients to all tissues, removes wastes, and transports gases, blood cells, immune cells, and hormones throughout the body.

Living up to their potential

The hematologic system manufactures new blood cells through a process called *hematopoiesis. Multipotential stem cells* in bone marrow give rise to five distinct cell types, called *unipotential stem cells.* Unipotential cells differentiate into one of the following four types of blood cells:
- erythrocyte (the most common type)
- granulocyte
- agranulocyte
- platelet.

(See *Tracing blood cell formation,* pages 142 and 143.)

I've got multipotential!

(Text continues on page 144)

Now I get it!

Tracing blood cell formation

Blood cells form and develop in the bone marrow by a process called *hematopoiesis*. This chart breaks down the process from when the five unipotential stem cells are "born" from the multipotential stem cell until they each reach "adulthood" as fully formed cells— erythrocytes, granulocytes, agranulocytes, or platelets.

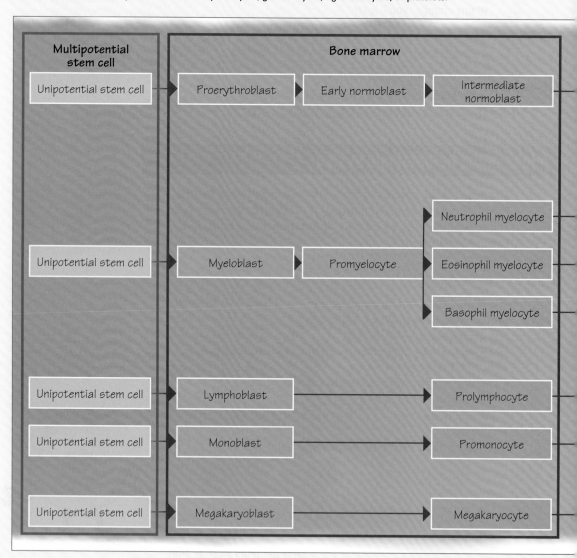

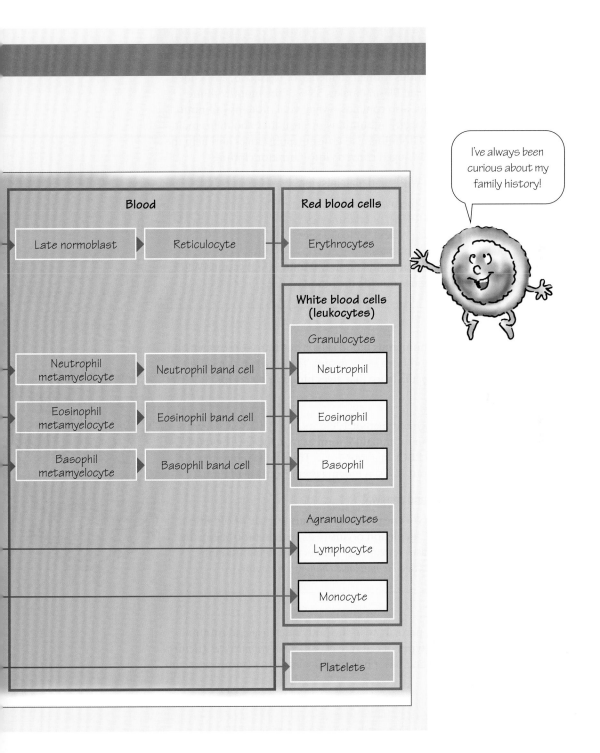

Blood

Late normoblast → Reticulocyte →

Red blood cells

Erythrocytes

White blood cells (leukocytes)

Granulocytes

Neutrophil metamyelocyte → Neutrophil band cell → Neutrophil

Eosinophil metamyelocyte → Eosinophil band cell → Eosinophil

Basophil metamyelocyte → Basophil band cell → Basophil

Agranulocytes

Lymphocyte

Monocyte

Platelets

I've always been curious about my family history!

Blood components

Blood consists of various formed elements, or *blood cells*, suspended in a fluid called *plasma*. Plasma makes 55% of the blood volume. It is mostly water, but also contains several vital proteins (albumin, globulin, and fibrinogen). Albumin is a transporter and helps maintain blood volume. Globulins also transport substances and help the immune system. Fibrinogen converts into fibrin for clotting.

The RBCs of blood—and the WBCs and platelets, too

Formed elements in the blood include:
- red blood cells (RBCs), or erythrocytes
- white blood cells (WBCs), or leukocytes
- platelets, or thrombocytes.

RBCs and platelets function entirely within blood vessels; WBCs act mainly in the tissues outside the blood vessels.

Red blood cells

RBCs are the most numerous blood cells. They transport oxygen and carbon dioxide to and from body tissues. They contain *hemoglobin*, the oxygen-carrying substance that gives blood its red color.

The life and times of the RBC

RBCs have an average life span of 120 days. Bone marrow releases RBCs into circulation in immature form as *reticulocytes*. The reticulocytes mature into RBCs in about 1 day. The spleen sequesters, or isolates, old, worn-out RBCs, removing them from circulation.

A balance between removal and renewal

The rate of reticulocyte release usually equals the rate of old RBC removal. When RBC depletion occurs (e.g., with hemorrhage), the bone marrow increases reticulocyte production to maintain the normal RBC count. (See *Hematologic changes with aging.*)

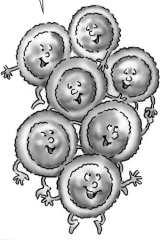

The body manufactures billions of new RBCs like us every day!

White blood cells

Five types of WBCs participate in the body's defense and immune systems. These five types of cells are classified as *granulocytes* (neutrophils, eosinophils, and basophils) and *agranulocytes* (monocytes and lymphocytes).

Granulocytes

Granulocytes include *neutrophils*, *eosinophils*, and *basophils*—collectively known as *polymorphonuclear leukocytes*. All granulocytes

Hematologic changes with aging

As a person ages, fatty bone marrow replaces some of the body's active blood-forming marrow—first in the long bones and later in the flat bones. The altered bone marrow can't increase erythrocyte production as readily in response to such stimuli as hormones, anoxia, hemorrhage, and hemolysis. Vitamin B_{12} absorption may also diminish with age, resulting in reduced erythrocyte mass and decreased hemoglobin levels and hematocrit.

contain a single *multilobular nucleus* and *granules* in the cytoplasm. Each cell type exhibits different properties, and each is activated by different stimuli.

Swallowing up your enemies

Neutrophils, the most numerous granulocytes, account for 55% to 65% of circulating WBCs. These phagocytic cells engulf, ingest, and digest foreign materials, such as bacteria and fungi. They leave the bloodstream by passing through the capillary walls into the tissues (a process called *diapedesis*) and then migrate to and accumulate at infection sites. Neutrophils are the first cells to arrive at the site of injury.

Neutrophils are the first cells to arrive at the site of injury.

Making the band

Worn-out neutrophils form the main component of pus. Bone marrow produces their replacements, immature neutrophils called *bands*. In response to infection, bone marrow must produce many immature cells and release them into circulation, elevating the band count.

Allies against allergies

Eosinophils account for 0.3% to 7% of circulating WBCs. These granulocytes also migrate from the bloodstream by diapedesis but do so as a response to an allergic reaction. Eosinophils accumulate in loose connective tissue, where they become involved in the ingestion of antigen-antibody complexes. Eosinophils also attack and kill parasitic organisms.

Memory jogger

To help yourself remember what **neutrophils** do, think of two other "n" words: numerous and neutralize.

Fighting the flames

Basophils usually constitute fewer than 2% of circulating WBCs. They possess little or no phagocytic ability. Their cytoplasmic granules secrete *histamine* in response to certain inflammatory and immune stimuli. Histamine makes the blood vessels more permeable and eases the passage of fluids from the capillaries into body tissues.

Agranulocytes

WBCs in the agranulocyte category—*monocytes* and *lymphocytes*—lack specific cytoplasmic granules and have nuclei without lobes. (See *Comparing granulocytes and agranulocytes*.)

We macrophages may be immobile at the moment, but at the first sign of inflammation, we're outta here!

The few and the large

Monocytes, the largest of the WBCs, constitute only 1% to 9% of WBCs in circulation. Like neutrophils, monocytes are phagocytic and enter the tissues by diapedesis. Outside the bloodstream, monocytes enlarge and mature, becoming tissue *macrophages* (also called *histiocytes*).

Protection against infection

As macrophages, monocytes may roam freely through the body when stimulated by inflammation. Usually, they remain immobile, populating most organs and tissues. Collectively, they serve as components of the *reticuloendothelial system*, which defends the body against infection and disposes of cell breakdown products.

Fluid finders

Macrophages concentrate in structures that filter large amounts of body fluid, such as the liver, spleen, and lymph nodes, where they defend against invading organisms. Macrophages are efficient *phagocytes*, cells that ingest microorganisms, cellular debris (including worn-out neutrophils), and necrotic tissue. When mobilized at an infection site, they phagocytize cellular remnants and promote wound healing.

Last and, in fact, least (in size)

Lymphocytes, the smallest of the WBCs and the second most numerous (20% to 43%), derive from stem cells in the bone marrow. There are three types of lymphocytes:
- *T lymphocytes* directly attack an infected cell. They mature in the **thymus**.
- *B lymphocytes* produce antibodies against specific antigens. They mature in the **bone** *marrow*.
- *Natural killer cells* provide immune surveillance and resistance to infection.

Now I get it!

Comparing granulocytes and agranulocytes

White blood cells (WBCs) are like soldiers fighting off the enemy. Each type of WBC fights a different enemy.

On the front line
Granulocytes, with "platoons" of basophils, neutrophils, and eosinophils, are the first forces "marshalled" against invading foreign organisms.

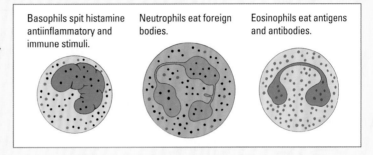

Basophils spit histamine antiinflammatory and immune stimuli.

Neutrophils eat foreign bodies.

Eosinophils eat antigens and antibodies.

In the trenches
Agranulocytes, with "platoons" of lymphocytes and monocytes, may roam freely on "patrol" when inflammation is reported, but they mainly "dig in" at structures that filter large amounts of fluid (such as the liver) and defend against invaders.

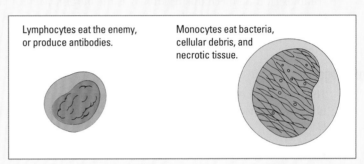

Lymphocytes eat the enemy, or produce antibodies.

Monocytes eat bacteria, cellular debris, and necrotic tissue.

Platelets

Platelets are small, colorless, disk-shaped cytoplasmic fragments split from cells in bone marrow called *megakaryocytes*.

These fragments, which have a life span of approximately 10 days, perform three vital functions:
- initiating contraction of damaged blood vessels to minimize blood loss
- forming *hemostatic plugs* in injured blood vessels
- with plasma, providing materials that accelerate blood coagulation.

Blood clotting

Hemostasis is the complex process by which *platelets, plasma,* and *coagulation factors* interact to control bleeding.

Stop the bleeding!

When a blood vessel ruptures, local *vasoconstriction* (decrease in the caliber of blood vessels) and *platelet clumping* (aggregation) at the site of the injury initially help prevent hemorrhage. The damaged cells then release tissue factor (thromboplastin), which activates the extrinsic pathway of the coagulation system.

A more long-term solution

However, formation of a more stable clot requires initiation of the complex clotting mechanisms known as the *intrinsic pathway*. This clotting system is activated by a protein, called factor XII, 1 of 12 substances necessary for coagulation and derived from plasma and tissue.

Come together

The final result of coagulation is a *fibrin clot,* an accumulation of a fibrous, insoluble protein at the site of the injury. (See *How blood clots.*)

When cells like me are damaged, we release thromboplastin, which activates the extrinsic portion of the coagulation system.

Coagulation factors

The materials that platelets and plasma provide work with *coagulation factors* to serve as *precursor compounds* in the clotting (coagulation) of blood.

12 factors clotting

Designated by name and Roman numeral, these *coagulation factors* are activated in a chain reaction, each one in turn activating the next factor in the chain:
- Factor I, *fibrinogen*, is a high-molecular-weight protein synthesized in the liver and converted to fibrin during the coagulation cascade.
- Factor II, *prothrombin*, is a protein synthesized in the liver in the presence of vitamin K and converted to thrombin during coagulation.

Now I get it!

How blood clots

When a blood vessel is severed or injured, three interrelated processes take place.

Constriction and aggregation

Immediately, the vessels affected by the injury contract (*constriction*), reducing blood flow. Also, platelets, stimulated by the exposed collagen of the damaged cells, begin to clump together (*aggregation*). Aggregation provides a temporary seal and a site for clotting to take place. The platelets release a number of substances that enhance constriction and aggregation.

Clotting pathways

Clotting, or *coagulation*—the transformation of blood from a liquid to a solid—may be initiated through two different pathways, the intrinsic pathway or the extrinsic pathway. The *intrinsic pathway* is activated when plasma comes in contact with damaged vessel surfaces. The *extrinsic pathway* is activated when tissue factor (a substance released by damaged endothelial cells) comes into contact with one of the clotting factors.

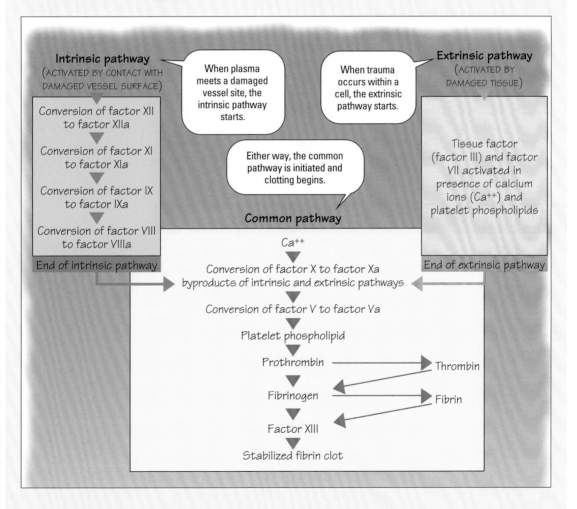

Intrinsic pathway (ACTIVATED BY CONTACT WITH DAMAGED VESSEL SURFACE)

When plasma meets a damaged vessel site, the intrinsic pathway starts.

When trauma occurs within a cell, the extrinsic pathway starts.

Extrinsic pathway (ACTIVATED BY DAMAGED TISSUE)

Conversion of factor XII to factor XIIa

Conversion of factor XI to factor XIa

Either way, the common pathway is initiated and clotting begins.

Conversion of factor IX to factor IXa

Tissue factor (factor III) and factor VII activated in presence of calcium ions (Ca++) and platelet phospholipids

Conversion of factor VIII to factor VIIIa

Common pathway

End of intrinsic pathway

Ca^{++}

End of extrinsic pathway

Conversion of factor X to factor Xa byproducts of intrinsic and extrinsic pathways

Conversion of factor V to factor Va

Platelet phospholipid

Prothrombin ———— Thrombin

Fibrinogen ———— Fibrin

Factor XIII

Stabilized fibrin clot

- Factor III, *tissue factor (thromboplastin)*, is released from damaged tissue; it's required to initiate the second phase, the extrinsic pathway.
- Factor IV, consisting of *calcium ions*, is required throughout the entire clotting sequence.
- Factor V, or *labile factor (proaccelerin)*, is a protein that's synthesized in the liver and functions during the common pathway phase of the coagulation system.
- Factor VII, *serum prothrombin conversion accelerator* or *stable factor (proconvertin)*, is a protein synthesized in the liver in the presence of vitamin K; it's activated by factor III in the extrinsic system.
- Factor VIII, *antihemophilic factor (antihemophilic globulin)*, is a protein synthesized in the liver and required during the intrinsic phase of the coagulation system.
- Factor IX, *plasma thromboplastin component*, a protein synthesized in the liver in the presence of vitamin K, is required in the intrinsic phase of the coagulation system.
- Factor X, *Stuart factor (Stuart-Prower factor)*, is a protein synthesized in the liver in the presence of vitamin K; it's required in the common pathway of the coagulation system.
- Factor XI, *plasma thromboplastin antecedent*, is a protein synthesized in the liver and required in the intrinsic pathway.
- Factor XII, *Hageman factor*, is a protein required in the intrinsic pathway.
- Factor XIII, *fibrin-stabilizing factor*, is a protein required to stabilize the fibrin strands in the common pathway phase of the coagulation system.

Blood groups

Blood groups are determined by the presence or absence of genetically determined *antigens* or *agglutinogens* (glycoproteins) on the surface of RBCs. A, B, and Rh are the most clinically significant blood antigens.

ABO groups

Testing for the presence of A and B antigens on RBCs is the most important system for classifying blood:
- Type A blood has A antigen on its surface.
- Type B blood has B antigen.

- Type AB blood has both A and B antigens.
- Type O blood has neither A nor B antigen.

Opposites don't attract

Plasma may contain *antibodies* that interact with these antigens, causing the cells to *agglutinate*, or combine into a mass. However, plasma can't contain antibodies to its own cell antigen or it would destroy itself. Thus, type A blood has A antigen but no A antibodies; however, it does have B antibodies.

Making a match

Precise blood typing and crossmatching (mixing and observing for agglutination of donor cells) are essential, especially for blood transfusions. A donor's blood must be compatible with a recipient's or the result can be fatal. The following blood groups are compatible:

- type A with type A or O
- type B with type B or O
- type AB with type A, B, AB, or O
- type O with type O only.

(See *Reviewing blood type compatibility*, page 152.)

Memory jogger

Blood types are easy to remember because they're named after the antigens they contain—**A** or **B** or both **A and B**—except for type **O**, which contains neither. The O serves as a nice visual reminder of that absence.

Rh typing

Rh typing determines whether Rh factor is present or absent in blood. Of the eight types of Rh antigens, only C, D, and E are common.

Positive and negative types

Typically, blood contains the Rh antigen. Blood with the Rh antigen is Rh positive; blood without the Rh antigen is Rh negative. Anti-Rh antibodies can appear only in a person who has become sensitized. Anti-Rh antibodies can appear in the blood of an Rh-negative person after entry of Rh-positive RBCs in the bloodstream—for example, from transfusion of Rh-positive blood. An Rh-negative female who carries an Rh-positive fetus may also acquire anti-Rh antibodies.

Now I get it!

Reviewing blood type compatibility

Precise blood typing and crossmatching can prevent the transfusion of incompatible blood, which can be fatal. Usually, typing the recipient's blood and crossmatching it with available donor blood take less than 1 hour.

Making a match

Agglutinogen (an antigen in red blood cells) and *agglutinin* (an antibody in plasma) distinguish the four ABO blood groups. This chart shows ABO compatibility from the perspectives of the recipient and the donor.

Blood group	Antibodies present in plasma	Compatible RBCs	Compatible plasma
RECIPIENT			
O	Anti-A and anti-B	O	O, A, B, AB
A	Anti-B	A, O	A, AB
B	Anti-A	B, O	B, AB
AB	Neither anti-A nor anti-B	AB, A, B, O	AB
DONOR			
O	Anti-A and anti-B	O, A, B, AB	O
A	Anti-B	A, AB	A, O
B	Anti-A	B, AB	B, O
AB	Neither anti-A nor anti-B	AB	AB, A, B, O

Blood typing is an important step before a transfusion.

Quick quiz

1. The component of blood that triggers defense and immune responses is the:
    A. WBC.
    B. platelet.
    C. RBC.
    D. hemoglobin.

Answer: A. Because of their phagocytic capabilities, WBCs serve as the body's first line of cellular defense against foreign organisms.

2.   The complex process by which platelets, plasma, and coagulation factors interact to control bleeding is called:
   A.   phagocytosis.
   B.   hematopoiesis.
   C.   hemostasis.
   D.   diapedesis.

Answer: C. Hemostasis is achieved through a three-part process: vasoconstriction, platelet aggregation, and coagulation.

3.   Blood cells form and develop in the:
   A.   platelet.
   B.   liver.
   C.   pancreas.
   D.   bone marrow.

Answer: D. Multipotential stem cells in the bone marrow give rise to five distinct cell types called unipotential stem cells. Each of these stem cells can differentiate into an erythrocyte, a granulocyte, an agranulocyte, or a platelet.

4.   The most numerous type of granulocytes are the:
   A.   bands.
   B.   neutrophils.
   C.   eosinophils.
   D.   basophils.

Answer: B. Neutrophils are the most numerous granulocytes, accounting for 50% to 75% of circulating WBCs.

5.   Blood groups are determined by testing for A and B antigens on the:
   A.   RBC.
   B.   WBC.
   C.   leukocyte.
   D.   platelet.

Answer: A. Blood groups are determined by the presence or absence of antigens or agglutinogens on the surface of RBCs.

Scoring

☆☆☆   If you answered all five questions correctly, wonderful! You're clearly thinking hemato-logically!

☆☆   If you answered four questions correctly, great. You've coagulated all the information in this chapter into a solid understanding of blood.

☆   If you answered fewer than four questions correctly, be positive (or A positive or AB positive). Just read the chapter again and give the quiz another go.

Just for fun!

Unscramble the words on the left to reveal the names of the three types of granulocytes. Then draw lines from each of the boxes to the lists on the right, linking each type to its particular characteristics

HENSLIPTOUR

_ _ _ _ _ _ _ _ _ _ _

LIESHOOPINS

_ _ _ _ _ _ _ _ _ _ _

ALBSHIPSO

_ _ _ _ _ _ _ _ _ _ _

A. Account for up to 7% of circulating WBCs
B. The most numerous of the granulocytes
C. Constitute fewer than 2% of circulating WBCs
D. Form main component of pus
E. Accumulate in loose connective tissue
F. Leave the bloodstream and migrate to and accumulate at infection sites
G. Possess little or no phagocytic ability
H. Engulf, ingest, and digest foreign materials
I. Migrate from the bloodstream as a response to an allergic reaction
J. Cytoplasmic granules secrete histamine
K. Replacements produced by bone marrow
L. Involved in the ingestion of antigen–antibody complexes

Answer: NEUTROPHILS: B, D, F, H, K EOSINOPHILS: A, E, I, L BASOPHILS: C, G, J

Selected Reference

Grossman, S. & Porth, C. M. (2014). *Porth's pathophysiology: Concepts of altered health states*. Philadelphia, PA: Lippincott Williams & Wilkins.

Chapter 10

Immune system

Just the facts

In this chapter, you'll learn:

♦ organs and tissues that make up the immune system
♦ functions of the immune system
♦ the body's response when the immune system fails.

A look at the immune system

The immune system defends the body against invasion by harmful organisms and chemical toxins.

Lymphoid rules

Organs and tissues of the immune system are referred to as "lymphoid" because they're all involved with the growth, development, and dissemination of lymphocytes, one type of white blood cell (WBC) also known as leukocytes. (See *Organs and tissues of the immune system*, page 156.)

The immune system has three major components:
- central lymphoid organs and tissue
- peripheral lymphoid organs and tissue
- accessory lymphoid organs and tissue.

Blood relatives

Although the immune system and blood are distinct entities, they're closely related. Their cells share a common origin in the bone marrow, and the immune system uses the bloodstream to transport its "troops" to the site of an invasion.

The immune system defends the body against invasion by harmful organisms and chemical toxins.

Body shop

Organs and tissues of the immune system

The immune system includes organs and tissues in which lymphocytes predominate as well as cells that circulate in the blood. This illustration shows central, peripheral, and accessory lymphoid organs and tissue.

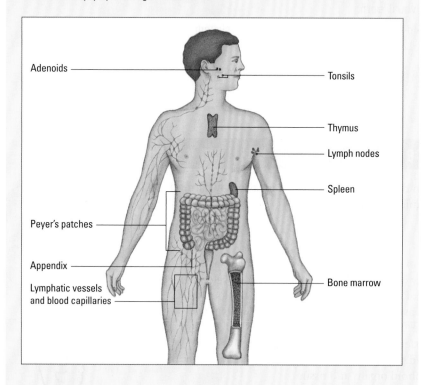

Adenoids

Tonsils

Thymus

Lymph nodes

Spleen

Peyer's patches

Appendix

Bone marrow

Lymphatic vessels and blood capillaries

We're T and B cells, the two major types of lymphocytes.

Central lymphoid organs and tissues

The bone marrow and the thymus each play a role in the development of B cells and T cells—the two major types of *lymphocytes*.

Bone marrow

The *bone marrow* contains stem cells, which can develop into any of several different cell types. Such cells are *multipotential*, meaning they're capable of taking many forms. The cells of the immune system and the blood develop from stem cells in a process called *hematopoiesis*.

To be B or to be T? That is the question...

Soon after their differentiation from other stem cells, some of the cells destined to become immune system cells serve as sources for *lymphocytes*; other cells of this differentiated group develop into *phagocytes* (cells that ingest microorganisms). Those that become lymphocytes are further differentiated to become either *B cells* (which mature in the bone marrow) or *T cells* (which travel to the thymus and mature there).

'Tis better to receive

B cells and T cells are distributed throughout the lymphoid organs, especially the lymph nodes and spleen. T and B lymphocytes have special receptors that respond to specific antigen molecule shapes. In B cells, this receptor is an immunoglobulin, also called an *antibody*. Antibodies attack pathogens or direct other cells, such as phagocytes, to attack for them.

I'm a multipotential cell. That means I'm capable of wearing many different hats.

Thymus

In fetuses and infants, the *thymus* is a two-lobed mass of lymphoid tissue that's located over the base of the heart in the mediastinum. The thymus helps form T lymphocytes (also known as *T cells*) for several months after birth. After this time, it has no function in the body's immunity. It reaches maximum size at puberty and then begins to atrophy until only a remnant remains in adults.

Basic training

In the thymus, T cells undergo a process called *T-cell education*, in which the cells are "trained" to recognize other cells from the same body (self cells) and distinguish them from all other cells (nonself cells). There are several types of T cells, each with a specific function:
- memory T cells
- helper T cells (T_4 cells)
- regulatory T cells (T_8 cells)
- natural killer T cells (cytotoxic T cells).

Memory jogger

To recall where lymphocytes mature, think, "B in B, and T in T." **B** cells mature in the **b**one marrow, and **T** cells mature in the **t**hymus.

Peripheral lymphoid organs and tissues

Peripheral structures of the immune system include the lymph nodes, the lymphatic vessels, and the spleen.

Lymph nodes

The *lymph nodes* are small, oval-shaped structures located along a network of *lymph channels*. Most abundant in the head, neck, axillae, abdomen, pelvis, and groin, lymph nodes help remove and destroy *antigens* (substances capable of triggering an immune response) that circulate in the blood and lymph.

Fully furnished compartments

Each lymph node is enclosed in a fibrous capsule. From this capsule, bands of connective tissue extend into the node and divide it into three compartments:
1. The *superficial cortex* contains follicles made up predominantly of B cells.
2. The *deep cortex* and interfollicular areas consist mostly of T cells.
3. The *medulla* contains numerous plasma cells that actively secrete *immunoglobulins*.

Lymphatic vessels

Lymph is a clear fluid that bathes the body tissues. It contains a liquid portion, which resembles blood plasma, as well as WBCs (mostly lymphocytes and macrophages) and antigens. Collected from body tissues, lymph seeps into *lymphatic vessels* across the vessels' thin walls. (See *Lymphatic vessels and lymph nodes*.)

Carried into the cavities…

Afferent lymphatic vessels carry lymph into the *subcapsular sinus* (or cavity) of the lymph node. From here, lymph flows through cortical sinuses and smaller radial medullary sinuses. Phagocytic cells in the deep cortex and medullary sinuses attack antigens carried in lymph. The antigens also may be trapped in the follicles of the superficial cortex. These processes essentially clean the lymph.

…and coming out cleansed

Cleansed lymph leaves the node through *efferent lymphatic vessels* at the *hilum* (a depression at the exit or entrance of the node). These vessels drain into *lymph node chains* that, in turn, empty into large lymph vessels, or trunks, which drain into the subclavian vein of the vascular system.

Getting security clearance

Usually, lymph travels through more than one lymph node because numerous nodes line the lymphatic channels that drain a particular region. For example, axillary nodes (located under the arm) filter drainage from the arms, and femoral nodes (in the inguinal region)

Memory jogger

Don't let "afferent" and "efferent" confuse you. Match the "a" in "**a**fferent" with the "a" in "**a**rrive" and the "e" in "**e**fferent" with the "e" in "**e**xit" so you'll know that lymph "**ar**-rives" in the sinuses through the "afferent" lymphatic vessels and "exits" the sinuses through the "efferent" lymphatic vessels.

Lymphatic vessels and lymph nodes

Lymphatic tissues are connected by a network of thin-walled drainage channels called *lymphatic vessels*. Resembling veins, the afferent lymphatic vessels carry lymph into lymph nodes; lymph slowly filters through the node and is collected into efferent lymphatic vessels.

It can check in but it can't check out
Lymphatic capillaries are located throughout most of the body. Wider than blood capillaries, they permit interstitial fluid to flow into them but not out.

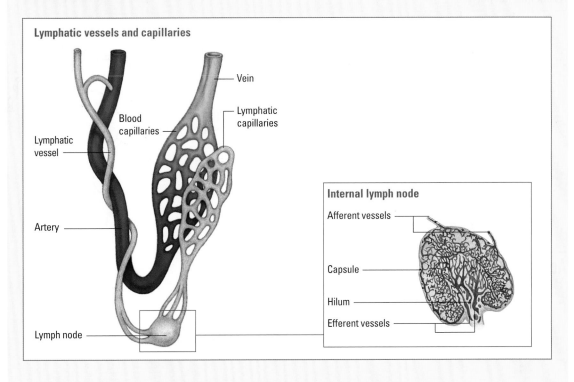

Lymphatic vessels and capillaries

- Vein
- Lymphatic capillaries
- Blood capillaries
- Lymphatic vessel
- Artery
- Lymph node

Internal lymph node

- Afferent vessels
- Capsule
- Hilum
- Efferent vessels

filter drainage from the legs. This arrangement prevents organisms that enter peripheral areas from migrating unchallenged to central areas.

Spleen

Located in the left upper quadrant of the abdomen beneath the diaphragm, the *spleen* is a dark red, oval structure that's approximately the size of a fist and is the largest lymphatic organ. Bands of connective tissue from the dense fibrous capsule surrounding the spleen extend into the spleen's interior.

The white and the red

The interior, called the *splenic pulp*, contains white and red pulp. *White pulp* contains compact masses of lymphocytes surrounding branches of the splenic artery. *Red pulp* consists of a network of blood-filled *sinusoids*, supported by a framework of reticular fibers and mononuclear phagocytes, along with some lymphocytes, plasma cells, and monocytes.

A real multitasker

The spleen has several functions:
- Its phagocytes engulf and break down worn-out red blood cells (RBCs), causing the release of hemoglobin, which then breaks down into its components. These phagocytes also selectively retain and destroy damaged or abnormal RBCs and cells with large amounts of abnormal hemoglobin.
- The spleen filters and removes bacteria and other foreign substances that enter the bloodstream; these substances are promptly removed by splenic phagocytes.
- Splenic phagocytes interact with lymphocytes to initiate an immune response.
- The spleen stores blood and 20% to 30% of platelets.
- If the spleen is removed due to disease or trauma, the liver and bone marrow assume its function.

The spleen is a dark red, oval structure that's approximately the size of a fist and is the largest lymphatic organ.

Accessory lymphoid organs and tissues

The *tonsils, adenoids, appendix,* and *Peyer's patches* remove foreign debris in much the same way lymph nodes do. They're located in areas in which microbial access is more likely, such as the nasopharynx (tonsils and adenoids) and the abdomen (appendix and Peyer's patches).

Immune system function

Immunity refers to the body's capacity to resist invading organisms and toxins, thereby preventing tissue and organ damage. The immune system is designed to recognize, respond to, and eliminate antigens, including bacteria, fungi, viruses, and parasites. It also preserves the body's internal environment by scavenging dead or damaged cells and patrolling for antigens.

Strategic moves

To perform these functions efficiently, the immune system uses three basic strategies:
- protective surface phenomena
- general host defenses
- specific immune responses.

Protective surface phenomena

Strategically placed physical, chemical, and mechanical barriers work to prevent the entry of potentially harmful organisms.

The forward guard

Intact and healing *skin* and *mucous membranes* provide the first line of defense against microbial invasion, preventing attachment of micro-organisms. Skin *desquamation* (normal cell turnover) and low pH further impede bacterial colonization. Seromucous surfaces are protected by antibacterial substances—for instance, the enzyme *lysozyme*, which is found in tears, saliva, and nasal secretions.

Breathe easy…

In the respiratory system (the easiest part of the body for microorganisms to enter), *nasal hairs* and *turbulent airflow* through the nostrils filter out foreign materials. Nasal secretions contain an immunoglobulin that discourages microbe adherence. Also, a mucous layer, which is continuously sloughed off and replaced, lines the respiratory tract and provides additional protection.

…and swallow

In the GI tract, bacteria are mechanically removed by saliva, swallowing, peristalsis, and defecation. In addition, the low pH of gastric secretions is *bactericidal* (bacteria-killing), rendering the stomach virtually free from live bacteria.

The remainder of the GI system is protected through *colonization resistance*, in which resident bacteria prevent other microorganisms from permanently making a home.

I'm a resident bacterium. I live in harmony with the body without causing disease, but I keep other microorganisms from colonizing my turf.

No colonization allowed

The urinary system is sterile except for the distal end of the urethra and the urinary meatus. Urine flow, low urine pH, immunoglobulin, and, in men, the bactericidal effects of *prostatic fluid* work together to impede bacterial colonization. A series of sphincters also inhibits bacterial migration.

General host defenses

When an antigen penetrates the skin or mucous membrane, the immune system launches nonspecific cellular responses in an effort to identify and remove the invader.

Raising the red flag

The first of the nonspecific responses against an antigen, the *inflammatory response*, involves vascular and cellular changes, including the production and release of such chemical substances as heparin, histamine, and kinin. These changes eliminate dead tissue, microorganisms, toxins, and inert foreign matter. (See *Understanding the inflammatory response.*)

Inflammatory response rousers

The following are leukocytes that are involved in the inflammatory response:

- *Neutrophils*, which are produced in the bone marrow, are the most numerous polymorphonuclear leukocytes Making up approximately 55% and 70% of the count of the WBC. They increase dramatically in number in response to infection and inflammation. They're the main constituent of pus and are highly mobile. Neutrophils are attracted to areas of inflammation. They engulf, digest, and dispose of invading organisms through a process called *phagocytosis*.
- *Eosinophils*, found in large numbers in the respiratory system and GI tract, multiply in allergic and parasitic disorders. The cell can be between 1% and 2% of the count of the WBC. Although their phagocytic function isn't clearly understood, evidence suggests that they participate in host defense against parasites.
- *Basophils* and *mast cells* also function in immune disorders and are approximately 1% of the count of the WBC. Basophils circulate in peripheral blood, whereas mast cells accumulate in connective tissue, particularly in the lungs, intestines, and skin. (Mast cells are not blood cells.) Both cells have surface receptors for immunoglobulin (Ig) E. When their receptors are cross-linked by an IgE antigen complex, they release mediators characteristic of the allergic response.

Specific immune responses

All foreign substances elicit the same general host defenses. In addition, particular microorganisms or molecules activate specific immune responses and can involve specialized sets of immune cells. Specific responses, classified as either *humoral immunity* or *cell-mediated immunity*, are produced by lymphocytes (B cells and T cells).

Now I get it!

Understanding the inflammatory response

The inflammatory response helps the body return to homeostasis after a wound occurs. Its primary function is to bring phagocytic cells (neutrophils and monocytes) to the inflamed area to destroy bacteria and rid the tissue spaces of dead and dying cells so that tissue repair can begin.

Inflammation produces five cardinal signs: redness, swelling, heat, pain, and decreased function. The first three signs result from local vasodilation, fluid leakage into the extravascular space, and blockage of lymphatic drainage. The fourth results from tissue space distention caused by swelling and pressure and from chemical irritation of nociceptors (pain receptors). The fifth sign results from tissue that could not repair itself and remains scar tissue.

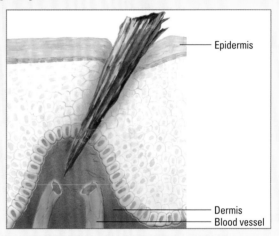

Epidermis

Dermis
Blood vessel

Humoral immunity

In this response, an invading antigen causes B cells to divide and differentiate into plasma cells. Each plasma cell, in turn, produces and secretes large amounts of antigen-specific immunoglobulins into the bloodstream.

The Ig guard

Each of the five types of *immunoglobulins* (IgA, IgD, IgE, IgG, and IgM) serves a particular function:
* IgA, IgG, and IgM guard against viral and bacterial invasion.
* IgD acts as an antigen receptor of B cells.
* IgE causes an allergic response.

"Y?" Because it's our job.

Immunoglobulins have a special molecular structure that creates a Y shape. The upper fork of the Y is designed to attach to a particular antigen; the lower stem enables the immunoglobulin to link with other structures in the immune system. Depending on the antigen, immunoglobulins can work in one of several ways:

My job is to produce antibodies. They attack the antigens for me.

- They can disable certain bacteria by linking with toxins that the bacteria produce; these immunoglobulins are called *antitoxins*.
- They can *opsonize* (coat) bacteria, making them targets for scavenging by phagocytosis. (See *How macrophages accomplish phagocytosis*.)
- Most commonly, they can link to antigens, causing the immune system to produce and circulate enzymes called *complement*.

I like the direct approach. Sometimes I attack antigens myself.

A break in the action

After the body's initial exposure to an antigen, a time lag occurs during which little or no antibody can be detected. During this time, the B cell recognizes the antigen and the sequence of division, differentiation, and antibody formation begins.

First response

The *primary antibody response* occurs 4 to 10 days after first-time antigen exposure. During this response, immunoglobulin levels increase and then quickly dissipate, and IgM antibodies form.

Second response: Hit 'em hard, hit 'em fast

Subsequent exposure to the same antigen initiates a *secondary antibody response*. In this response, memory B cells manufacture antibodies (now mainly IgG), achieving peak levels in 1 to 2 days. These elevated levels persist for months and then fall slowly. Thus, the secondary antibody response is faster, more intense, and more persistent than is the primary response. This response intensifies with each subsequent exposure to the same antigen.

Getting complex

After the antibody reacts to the antigen, an *antigen-antibody complex* forms. The complex serves several functions. First, a macrophage processes the antigen and presents it to antigen-specific B cells. Then the antibody activates the complement system, causing an *enzymatic cascade* that destroys the antigen.

You thought I was slow in the first round? Well, just wait. I always come back faster and meaner in the second.

Complement system

The *complement system* is activated by a tissue injury or antigen-antibody reactions. It bridges humoral and cell-mediated immunity and attracts phagocytic neutrophils and macrophages to the antigen site.

Working together

Indispensable to the humoral immune response, the complement system consists of about 25 enzymes that "complement" the work of antibodies by aiding phagocytosis or destroying bacteria cells (through puncture of their cell membranes).

Now I get it!

How macrophages accomplish phagocytosis

Microorganisms and other antigens that invade the skin and mucous membranes are removed by *phagocytosis*, a defense mechanism carried out by macrophages (mononuclear leukocytes) and neutrophils (polymorphonuclear leukocytes). Here's how macrophages accomplish phagocytosis.

Chemotaxis
Chemotactic factors attract macrophages to the antigen site.

Microorganism

Chemotactic factors

Macrophage

Opsonization
The antibody (immunoglobulin G) or complement fragment coats the microorganism, enhancing macrophage binding to the antigen, now called an opsinogen.

Opsonized microorganism

Ingestion
The macrophage extends its membrane around the opsonized microorganism, engulfing it within a vacuole (phagosome).

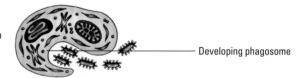

Developing phagosome

Digestion
As the phagosome shifts away from the cell periphery, it merges with lysosomes, forming a phagolysosome, where antigen destruction occurs.

Phagolysosome

Release
When digestion is complete, the macrophage expels digestive debris, including lysosomes, prostaglandins, complement components, and interferon, which continue to mediate the immune response.

Digestive debris

A ripple effect

Complement proteins travel in the bloodstream in an inactive form. When the first complement substance is triggered (typically by an antibody interlocked with an antigen), it sets in motion a ripple effect. As each component is activated in turn, it acts on the next component in a sequence of carefully controlled steps called the *complement cascade*.

Attack mode

This cascade leads to the creation of the *membrane attack complex*. Inserted into the membrane of the target cell, this complex creates a channel through which fluids and molecules flow in and out. The target cell then swells and eventually bursts.

Other benefits flow from the complement cascade

By-products of the complement cascade also produce:
- the inflammatory response (resulting from release of the contents of mast cells and basophils)
- stimulation and attraction of neutrophils (which participate in phagocytosis)
- coating of target cells by C3b (an inactivated fragment of the complement protein C3), making them attractive to phagocytes.

Cell-mediated immunity

Cell-mediated immunity protects the body against bacterial, viral, and fungal infections by inactivating the antigen and provides resistance against transplanted cells and tumor cells.

Ever vigilant

In this immune response, a macrophage processes the antigen, which is then presented to T cells. Some T cells become sensitized and destroy the antigen; others release *lymphokines*, which activate macrophages that destroy the antigen. Sensitized T cells then travel through the blood and lymphatic systems, providing ongoing surveillance in their quest for specific antigens. (See *Immune response to bacterial invasion*.)

The great communicators

Cytokines are low-molecular-weight proteins involved in the communication between macrophages and lymphocytes. These proteins are responsible for inducing and regulating many immune and inflammatory responses. Cytokines include colony-stimulating factors, interferons, interleukins, tumor necrosis factors, and transforming growth factor. They're an important part of a well-functioning immune system.

The complement cascade plays a crucial role in the inflammatory response.

Now I get it!

Immune response to bacterial invasion

Invasion of a foreign substance can trigger two types of immune responses—antibody-mediated (humoral) and cell-mediated immunity:

• In *humoral* immunity, antigens stimulate B cells to differentiate into plasma cells and produce circulating antibodies that disable bacteria and viruses before they can enter host cells.

• In *cell-mediated* immunity, T cells move directly to attack invaders. Three T-cell subgroups trigger the response to infection. Helper T cells spur B cells to manufacture antibodies. Effector T cells kill antigens and produce lymphokines (proteins that induce the inflammatory response and mediate the delayed hypersensitivity reaction). Suppressor T cells regulate T and B types of immune response.

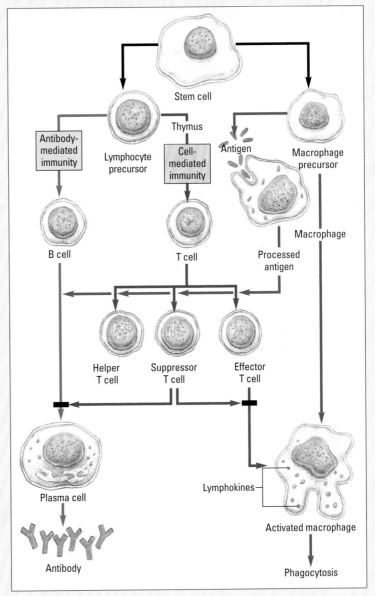

Immune system malfunction

Because of their complexity, the processes involved in host defense and immune response may malfunction. When the body's defenses are exaggerated, misdirected, or either absent or depressed, the result may be a *hypersensitivity disorder*, *autoimmunity*, or *immunodeficiency*, respectively.

Hypersensitivity disorders

An exaggerated or inappropriate immune response may lead to various hypersensitivity disorders.

Typing them out

Such disorders are classified as type I through type IV, depending on which immune system activity causes tissue damage, although some overlap exists:

- Type I disorders are *anaphylactic (immediate, atopic, IgE-mediated reaginic) reactions*. Examples of type I disorders include systemic anaphylaxis, hay fever (seasonal allergic rhinitis), reactions to insect stings, some food and drug reactions, some cases of urticaria, and infantile eczema.
- Type II disorders are *cytotoxic (cytolytic, complement-dependent cytotoxicity) reactions*. Examples of type II disorders include Goodpasture's syndrome, autoimmune hemolytic anemia, transfusion reactions, hemolytic disease of the neonate, myasthenia gravis, and some drug reactions.
- Type III disorders are *immune complex disease reactions*. Examples of type III disorders are reactions associated with such infections as hepatitis B and bacterial endocarditis; cancers, in which a serum sickness-like syndrome may occur; and autoimmune disorders such as systemic lupus erythematosus. This hypersensitivity reaction may also follow drug or serum therapy.
- Type IV disorders are *delayed (cell-mediated) hypersensitivity reactions*. Type IV disorders include tuberculin reactions, contact hypersensitivity (latex allergy), and sarcoidosis.

Hypersensitivity stems from an exaggerated or inappropriate immune response.

Autoimmune disorders

Autoimmune disorders are marked by an abnormal response to one's own tissue.

Diffusion can lead to confusion

Autoimmunity leads to a sequence of tissue reactions and damage that may produce diffuse systemic signs and symptoms. Among the autoimmune disorders are type 1 diabetes mellitus, rheumatoid arthritis, juvenile rheumatoid arthritis, psoriatic arthritis, ankylosing spondylitis, Sjögren's syndrome, multiple sclerosis, autoimmune pancreatitis, and lupus erythematosus. (See *Immunologic changes with aging*.)

Immunodeficiency disorders occur when the immune system is depressed or on a downward slide.

Immunodeficiency

Immunodeficiency disorders are caused by an absent or a depressed immune response in various forms.

Unfortunately, no deficiency of immunodeficiency disorders

Immunodeficiency disorders include X-linked infantile hypogammaglobulinemia, common variable immunodeficiency, DiGeorge syndrome, acquired immunodeficiency syndrome, chronic granulomatous disease, ataxia-telangiectasia, severe combined immunodeficiency disease, and complement deficiencies.

 Senior moment

Immunologic changes with aging

Immune function starts declining at sexual maturity and continues declining with age. During this decline, the immune system begins losing its ability to differentiate between self and nonself, and the incidence of autoimmune disease increases. The immune system also begins losing its ability to recognize and destroy mutant cells, which presumably accounts for the increase in cancer among older people.

External factors, such as nutritional status and exposure to chemical and environmental pollution and ultraviolet radiation, can also affect immune status.

Decreased antibody response in older people makes them more susceptible to infection. Tonsillar atrophy and lymphadenopathy commonly occur.

Total and differential leukocyte counts don't change significantly with age. However, some people over age 65 may exhibit a slight decrease in leukocyte count. When this happens, the number of B cells and total lymphocytes decreases, and T cells decrease in number and become less effective. Also, the sizes of the lymph nodes and spleen reduce slightly.

Quick quiz

1. Stem cells are multipotential and develop into other types of cells through which of the following processes?
    A. The process of chemotaxis
    B. The process of phagocytosis
    C. The process of hematopoiesis
    D. The process of opsonization

Answer: C. Hematopoiesis is the formation of blood cells, which occurs in the bone marrow.

2. When the lymph is cleansed, it exits the lymph nodes through which of the following routes?
    A. The blood capillaries
    B. The lymphatic capillaries
    C. The afferent lymphatic vessels
    D. The efferent lymphatic vessels

Answer: D. Efferent lymphatic vessels drain into lymph node chains, then into large lymph vessels, and, finally, into the subclavian vein.

3. During which phase of phagocytosis does a macrophage engulf an opsonized microorganism within a vacuole?
    A. Chemotaxis
    B. Opsonization
    C. Ingestion
    D. Digestion

Answer: C. During ingestion, the macrophage extends its membrane around the microorganism, engulfing it within a vacuole and forming a phagosome.

Scoring

☆☆☆   If you answered all three questions correctly, impressive! You've responded like a pro to all of our immunologic challenges!

☆☆   If you answered two questions correctly, prodigious! You've proved your multipotential!

☆   If you answered only one question correctly, you might say you have an immunodeficiency. Take some vitamin C, and read this chapter again in the morning!

Selected References

Ignatavicius, D. D. & Workman, M. L., (2016). *Medical-surgical nursing: Patient-Centered collaborative care* (8th ed.). St. Louis, MO: Elsevier.

Jarvis, C. (2012). *Physical examination & health assessment* (6th ed.). St. Louis, MO: Elsevier.

Chapter 11

Respiratory system

Just the facts

In this chapter, you'll learn:

♦ structures of the respiratory system and their functions

♦ the processes of inspiration and expiration

♦ the way in which gas exchange takes place

♦ problems with the nervous, musculoskeletal, and pulmonary systems that can affect breathing

♦ the role of the lungs in acid-base balance.

A look at the respiratory system

The respiratory system maintains the exchange of oxygen and carbon dioxide in the lungs and tissues. It also helps regulate the body's acid-base balance. Functionally, the respiratory system is composed of a conducting zone and a respiratory zone. The conducting zone consists of the continuous passageway that transports air in and out of the lungs (nose, pharynx, larynx, trachea, bronchi, and bronchioles). The respiratory zone, composed of the bronchioles, alveolar ducts, and alveoli, performs gas exchange.

The respiratory system consists of the:

- upper respiratory tract
- lower respiratory tract
- thoracic cavity.

Hey, want to trade? I'll give you a little oxygen for a bit of carbon dioxide.

Upper respiratory tract

The upper respiratory tract consists primarily of the nose (nostrils and nasal passages), mouth, nasopharynx, oropharynx, laryngopharynx, and larynx. These structures filter, warm, and humidify inspired air.

They're also responsible for detecting taste and smell and chewing and swallowing food. (See *Structures of the respiratory system*.)

Nostrils and nasal passages

Air enters the body through the nostrils (*nares*). In the nares, small hairs known as *vibrissae* filter out foreign particles such as dust. Air then passes into the two nasal passages, which are separated by the *septum*. Cartilage forms the anterior walls of the nasal passages; bony structures (*conchae* or *turbinates*) form the posterior walls.

Just passing through

The *conchae* warm and humidify air before it passes into the nasopharynx. Their mucus layer also traps finer foreign particles, which the *cilia* (small, hairlike projections) carry to the pharynx to be swallowed.

It's nothing to sneeze at. These involuntary defense mechanisms help protect the respiratory system from infection and foreign body inhalation.

Sinuses and nasopharynx

The four paranasal sinuses are located in the frontal, sphenoid, and maxillary bones. The sinuses provide speech resonance.

Air passes from the nasal cavity into the muscular nasopharynx through the *choanae,* a pair of posterior openings in the nasal cavity that remain constantly open. The nasopharynx is located behind the nose and above the throat.

Oropharynx and laryngopharynx

The oropharynx is the posterior wall of the mouth. It connects the nasopharynx and the laryngopharynx. The laryngopharynx extends to the esophagus and larynx.

Larynx

The larynx contains the vocal cords and connects the pharynx with the trachea, part of the lower respiratory tract. Muscles and cartilage form the walls of the larynx, including the large, shield-shaped thyroid cartilage situated just under the jaw line.

Structures of the respiratory system

This illustration depicts the structures of the respiratory system, which can be divided into the upper and lower respiratory tracts.

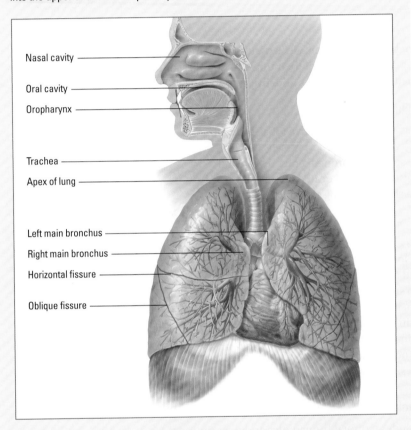

Nasal cavity

Oral cavity

Oropharynx

Trachea

Apex of lung

Left main bronchus

Right main bronchus

Horizontal fissure

Oblique fissure

Lower respiratory tract

The lower respiratory tract consists of the trachea, bronchi, and lungs. These structures contain a mucous membrane with hairlike cilia that lines the lower tract. Cilia constantly clean the tract and carry foreign matter upward for swallowing or expectoration.

Trachea

The trachea extends from the *cricoid cartilage* at the top to the carina (also called the *tracheal bifurcation*). C-shaped cartilage rings reinforce and protect the trachea to prevent it from collapsing. The carina is a ridge-shaped structure at the level of T6 or T7.

Bronchi

The primary bronchi begin at the carina. The right mainstem bronchus—shorter, wider, and more vertical than the left—supplies air to the right lung. The left mainstem bronchus delivers air to the left lung.

> Bronchi branch out—from lobar bronchi to segmental bronchi to bronchioles.

Secondary bronchi and the hilum

The mainstem bronchi divide into the five lobar bronchi (secondary bronchi). Along with blood vessels, nerves, and lymphatics, the secondary bronchi enter the pleural cavities and the lungs at the *hila*. Located behind the heart, the hilum is a slit on the lung's medial surface.

Branching out

One lobar bronchus enters a lobe in each lung. Within its lobe, each of the lobar bronchi branches into segmental bronchi (tertiary bronchi). The segments continue to branch into smaller and smaller bronchi, finally branching into bronchioles.

The larger bronchi consist of cartilage, smooth muscle, and epithelium. As the bronchi become smaller, they lose cartilage and then smooth muscle. Ultimately, the smallest bronchioles consist of just a single layer of epithelial cells.

Respiratory bronchioles

Each bronchiole includes terminal bronchioles and the acinus—the chief respiratory unit for gas exchange. (See *A close look at a pulmonary airway*.)

Within the acinus, terminal bronchioles branch into yet smaller respiratory bronchioles. The respiratory bronchioles feed directly into alveoli at sites along their walls.

A close look at a pulmonary airway

The respiratory unit consists of the respiratory bronchiole, alveolar duct and sac, and alveoli. Gas exchange occurs rapidly in the alveoli, in which oxygen from inhaled air diffuses into the blood and carbon dioxide diffuses from the blood into exhaled air.

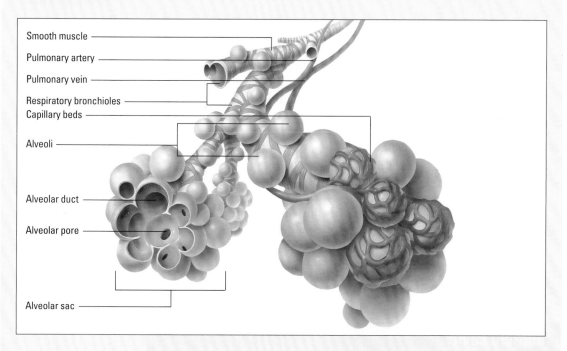

Smooth muscle
Pulmonary artery
Pulmonary vein
Respiratory bronchioles
Capillary beds
Alveoli
Alveolar duct
Alveolar pore
Alveolar sac

Alveoli

The respiratory bronchioles eventually become alveolar ducts, which terminate in clusters of capillary-swathed alveoli called *alveolar sacs.* Gas exchange takes place through the alveoli.

Alveolar walls contain two basic epithelial cell types:
1. Type I cells are the most abundant. It is across these thin, flat, squamous cells that gas exchange occurs.
2. Type II cells secrete *surfactant,* a substance that coats the alveolus and promotes gas exchange by lowering surface tension and preventing lung collapse.

Lungs

The cone-shaped lungs hang suspended in the right and left pleural cavities, straddling the heart, and anchored by root and pulmonary ligaments. The right lung is shorter, broader, and larger than the left. It has three lobes and handles 55% of gas exchange. The left lung has two lobes. Each lung's concave base rests on the diaphragm; the apex extends about ½″ (1.5 cm) above the first rib.

Pleura and pleural cavities

The pleura—the membrane that totally encloses the lung—is composed of a visceral layer and a parietal layer. The visceral pleura hugs the entire lung surface, including the areas between the lobes. The parietal pleura lines the inner surface of the chest wall and upper surface of the diaphragm.

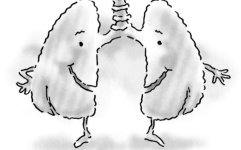

We hang out in the right and left pleural cavities.

Serous fluid has serious functions

The pleural cavity—the tiny area between the visceral and parietal pleural layers—contains a thin film of serous fluid. This fluid has two functions:
1. It lubricates the pleural surfaces, which allows them to slide smoothly against each other as the lungs expand and contract.
2. It creates a bond between the layers that causes the lungs to move with the chest wall during breathing.

Thoracic cavity

The thoracic cavity contains the lungs. It is surrounded by the diaphragm (below), the scalene muscles and fasciae of the neck (above), and the ribs, intercostal muscles, vertebrae, sternum, and ligaments (around the circumference).

Mediastinum

The space between the lungs is called the *mediastinum*. It contains the:
- heart and pericardium
- thoracic aorta
- pulmonary artery and veins
- venae cavae and azygos veins

- thymus, lymph nodes, and vessels
- trachea, esophagus, and thoracic duct
- vagus, cardiac, and phrenic nerves.

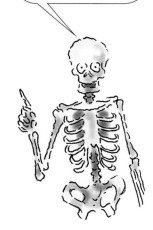

The ribs, like the vertebrae, are numbered from top to bottom.

Thoracic cage

The thoracic cage is composed of bone and cartilage. It supports and protects the lungs, allowing them to expand and contract.

Posterior thoracic cage

The vertebral column and 12 pairs of ribs form the posterior portion of the thoracic cage. The ribs form the major portion of the thoracic cage. They extend from the thoracic vertebrae toward the anterior thorax.

Anterior thoracic cage

The anterior thoracic cage consists of the manubrium, sternum, xiphoid process, and ribs. It protects the mediastinal organs that lie between the right and left pleural cavities.

Attached or floating free?

Ribs 1 through 7 attach directly to the sternum; ribs 8 through 10 attach to the cartilage of the preceding rib. The other 2 pairs of ribs are "free-floating"—they don't attach to any part of the anterior thoracic cage. Rib 11 ends anterolaterally, and rib 12 ends laterally.

Bordering on the costal angle

The lower parts of the rib cage (costal margins) near the xiphoid process form the borders of the costal angle—an angle of about 90 degrees in a normal person. (See *Locating lung structures in the thoracic cage*, page 178.)

It's suprasternal

Above the anterior thorax is a depression called the suprasternal notch. Because the suprasternal notch isn't covered by the rib cage like the rest of the thorax, the trachea and aortic pulsation can be palpated here. (See *Respiratory changes with aging*, page 179.)

Because the suprasternal notch isn't covered by the rib cage, the trachea and aortic pulsation can be palpated here.

Body shop

Locating lung structures in the thoracic cage

The ribs, vertebrae, and other structures of the thoracic cage act as landmarks that can be used to identify underlying structures.

From an anterior view
• The base of each lung rests at the level of the sixth rib at the midclavicular line and the eighth rib at the midaxillary line.
• The apex of each lung extends about ¾" to 1½" (2 to 4 cm) above the medial aspect of the clavicles.
• The upper lobe of the right lung ends level with the fourth rib at the midclavicular line and with the fifth rib at the midaxillary line.
• The middle lobe of the right lung extends triangularly from the fourth to the sixth rib at the midclavicular line and to the fifth rib at the midaxillary line.
• Because the left lung doesn't have a middle lobe, the upper lobe of the left lung ends level with the fourth rib at the midclavicular line and with the fifth rib at the midaxillary line.

From a posterior view
• The lungs extend from the cervical area to the level of T10. On deep inspiration, the lungs may descend to T12.
• An imaginary line, stretching from the T3 level along the inferior border of the scapulae to the fifth rib at the midaxillary line, separates the upper lobes of both lungs.
• The upper lobes are situated above T3; the lower lobes are situated below T3 and extend to the level of T10.
• The diaphragm originates around the ninth or tenth rib.

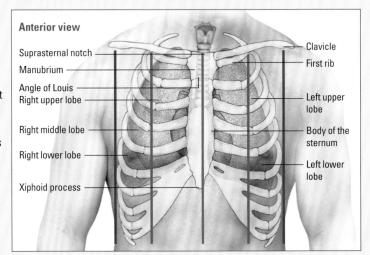

Anterior view

Suprasternal notch
Manubrium
Angle of Louis
Right upper lobe
Right middle lobe
Right lower lobe
Xiphoid process

Clavicle
First rib
Left upper lobe
Body of the sternum
Left lower lobe

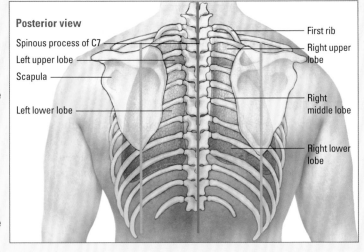

Posterior view

Spinous process of C7
Left upper lobe
Scapula
Left lower lobe

First rib
Right upper lobe
Right middle lobe
Right lower lobe

Respiratory changes with aging

As a person ages, the body undergoes respiratory system changes. These changes may be both structural and functional in nature.

Structural changes

Age-related anatomic changes in the upper airways include nose enlargement from continued cartilage growth, general atrophy of the tonsils, and tracheal deviations from changes in the aging spine. Possible thoracic changes include increased anteroposterior chest diameter (resulting from altered calcium metabolism) and calcification of costal cartilage, which reduces mobility of the chest wall. Kyphosis advances with age because of such factors as osteoporosis and vertebral fractures.

The lungs become more rigid and the number and size of alveoli decline with age.

Pulmonary function changes

Pulmonary function decreases in older people as a result of respiratory muscle degeneration or atrophy. Ventilatory capacity diminishes for several reasons:
• The lungs' diffusing capacity declines; decreased inspiratory and expiratory muscle strength diminishes vital capacity.
• Lung tissue degeneration causes a decrease in the lungs' elastic recoil capability, which results in an elevated residual volume.
• Closing of some airways produces poor ventilation of the basal areas, resulting in both a decreased surface area for gas exchange and reduced partial pressure of oxygen. Oxygen saturation decreases and may lead to decreased exercise tolerance.

Inspiration and expiration

Breathing involves two actions: inspiration (an active process) and expiration (a relatively passive process). Both actions rely on respiratory muscle function and the effects of pressure differences in the lungs.

It's perfectly normal!

During normal respiration, the external intercostal muscles aid the diaphragm, the major muscle of respiration. The diaphragm descends to lengthen the chest cavity, while the external intercostal muscles (located between and along the lower borders of the ribs) contract to expand the anteroposterior diameter. This coordinated action causes a reduction in intrapleural pressure, and inspiration occurs. Rising of the diaphragm and relaxation of the intercostal muscles cause an increase in intrapleural pressure, and expiration results. (See *Muscles of respiration*, page 180.)

Breathing involves two actions: inspiration and expiration.

Body shop

Muscles of respiration

The muscles of respiration help the chest cavity expand and contract. Pressure differences between atmospheric air and the lungs help produce air movement. These illustrations show the muscles that work together to allow inspiration and expiration.

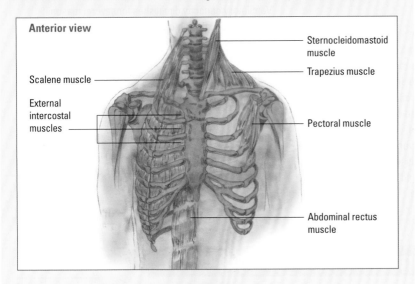

Anterior view

Sternocleidomastoid muscle

Trapezius muscle

Scalene muscle

External intercostal muscles

Pectoral muscle

Abdominal rectus muscle

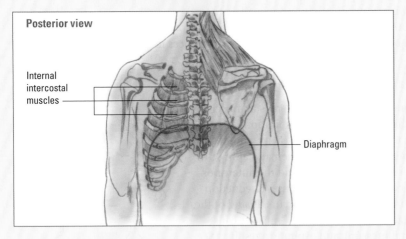

Posterior view

Internal intercostal muscles

Diaphragm

Forced inspiration and active expiration

During exercise, when the body needs increased oxygenation, or in certain disease states that require forced inspiration and active expiration, the accessory muscles of respiration also participate.

Forced inspiration

During forced inspiration:

- the pectoral muscles in the upper chest raise the chest to increase the anteroposterior diameter
- the sternocleidomastoid muscles in the side of the neck raise the sternum
- the scalene muscles in the neck elevate, fix, and expand the upper chest
- the posterior trapezius muscles in the upper back raise the thoracic cage.

Active expiration

During active expiration, the internal intercostal muscles contract to shorten the chest's transverse diameter and the abdominal rectus muscles pull down the lower chest, thus depressing the lower ribs. (See *Mechanics of ventilation*, page 182.)

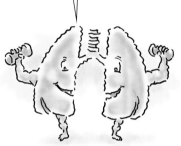

When the body's demand for oxygen is increased, such as during exercise, the accessory muscles of respiration help us out.

Gas station

Oxygen-depleted blood enters the lungs from the pulmonary artery of the heart's right ventricle and then flows through the main pulmonary arteries into the smaller vessels of the pleural cavities and the main bronchi, through the arterioles and, eventually, to the capillary networks in the alveoli. Gas exchange—oxygen and carbon dioxide diffusion—takes place in the alveoli.

Internal and external respiration

Effective respiration consists of gas exchange in the lungs, called *external respiration,* and gas exchange in the tissues, called *internal respiration.*

Internal respiration occurs only through diffusion. External respiration occurs through three processes:

1. *ventilation*—gas distribution into and out of the pulmonary airways

Now I get it!

Mechanics of ventilation

Breathing results from differences between atmospheric and intrapulmonary pressures, as described below.

1. Before inspiration, intrapulmonary pressure equals atmospheric pressure (approximately 760 mm Hg). Intrapleural pressure is 756 mm Hg.

2. The intrapulmonary atmospheric pressure gradient pulls air into the lungs until the two pressures are equal.

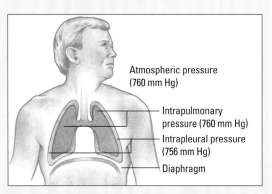

Atmospheric pressure (760 mm Hg)

Intrapulmonary pressure (760 mm Hg)

Intrapleural pressure (756 mm Hg)

Diaphragm

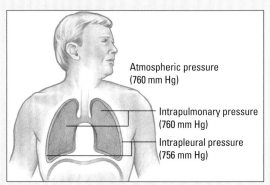

Atmospheric pressure (760 mm Hg)

Intrapulmonary pressure (760 mm Hg)

Intrapleural pressure (756 mm Hg)

3. During inspiration, the diaphragm and external intercostal muscles contract, enlarging the thorax vertically and horizontally. As the thorax expands, intrapleural pressure decreases and the lungs expand to fill the enlarging thoracic cavity.

4. During normal expiration, the diaphragm slowly relaxes and the lungs and thorax passively return to resting size and position. During deep or forced expiration, contraction of internal intercostal and abdominal muscles reduces thoracic volume. Lung and thorax compression raises intrapulmonary pressure above atmospheric pressure.

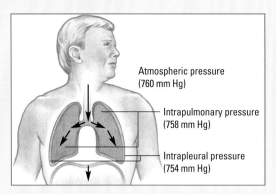

Atmospheric pressure (760 mm Hg)

Intrapulmonary pressure (758 mm Hg)

Intrapleural pressure (754 mm Hg)

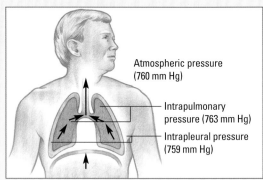

Atmospheric pressure (760 mm Hg)

Intrapulmonary pressure (763 mm Hg)

Intrapleural pressure (759 mm Hg)

2. *pulmonary perfusion*—blood flow from the right side of the heart, through the pulmonary circulation, and into the left side of the heart
3. *diffusion*—gas movement through a semipermeable membrane from an area of greater concentration to one of lesser concentration.

Ventilation

Ventilation is the distribution of gases (oxygen and carbon dioxide) into and out of the pulmonary airways. Problems within the nervous, musculoskeletal, and pulmonary systems greatly compromise breathing effectiveness.

The central nervous system's respiratory center is located in the lateral medulla.

Nervous system influence

Involuntary breathing results from stimulation of the respiratory center in the medulla and the pons of the brain. Central chemical receptors in the medulla indirectly monitor the level of carbon dioxide in the blood. Carbon dioxide exerts the main influence on breathing. When carbon dioxide levels rise, the rate and depth of breathing increase to eliminate excess carbon dioxide.

Peripheral chemical receptors in the aorta and carotid arteries monitor the level of oxygen in the blood. When oxygen levels drop, respiratory rate and depth increase to improve the blood oxygen level. However, the peripheral chemical receptors are less sensitive than the central receptors and don't respond until oxygen levels are quite low.

Musculoskeletal influence

The adult thorax is flexible—its shape can be changed by contracting the chest muscles, thus expanding the volume of the thoracic cavity. The medulla controls ventilation primarily by stimulating contraction of the diaphragm and external intercostal muscles. These actions produce the intrapulmonary pressure changes that cause inspiration.

Pulmonary influence

Airflow distribution can be affected by many factors:
- airflow pattern (see *Comparing airflow patterns*, page 184)
- volume and location of the functional reserve capacity (air retained in the alveoli that prevents their collapse during expiration)
- degree of intrapulmonary resistance
- presence of lung disease.

The path of least resistance

If airflow is disrupted for any reason, airflow distribution follows the path of least resistance.

Comparing airflow patterns

The pattern of airflow through the respiratory passages affects airway resistance.

Laminar flow
Laminar flow, a linear pattern that occurs at low flow rates, offers minimal resistance. This flow type occurs mainly in the small peripheral airways of the bronchial tree.

Turbulent flow
The eddying pattern of turbulent flow creates friction and increases resistance. Turbulent flow is normal in the trachea and large central bronchi. If the smaller airways become constricted or clogged with secretions, however, turbulent flow may also occur there.

Transitional flow
A mixed pattern known as transitional flow is common at lower flow rates in the larger airways, especially where the airways narrow from obstruction, meet, or branch.

Increased workload, decreased efficiency

Other musculoskeletal and intrapulmonary factors can affect airflow and, in turn, may affect breathing. For instance, forced breathing (as occurs in emphysema) activates accessory muscles of respiration, which require additional oxygen to work. This results in less efficient ventilation with an increased workload.

Airflow interference and alterations

Other airflow alterations can also increase oxygen and energy demand and cause respiratory muscle fatigue. These conditions include interference with expansion of the lungs or thorax (changes in compliance) and interference with airflow in the tracheobronchial tree (changes in resistance). Both can result in reduced tidal volume and alveolar ventilation.

Pulmonary perfusion

Pulmonary perfusion refers to blood flow from the right side of the heart, through the pulmonary circulation, and into the left side of the heart. Perfusion aids external respiration. Normal pulmonary blood flow allows alveolar gas exchange, and multiple factors are involved in ensuring adequate perfusion:

- pressure within the alveoli
- gravitational forces
- pressure within both arteries and veins.

Ventilation–perfusion match

Ventilation and perfusion differ in the various parts of the lungs. For example, ventilation is greater in areas that are perfused better, such as the lung bases. The tops of the lungs, or apices, require less perfusion and, therefore, also less ventilation. Areas in which perfusion and ventilation are similar have what is referred to as a *ventilation-perfusion match;* in such areas, gas exchange is most efficient. (See *What happens in ventilation-perfusion mismatch,* page 186.)

During diffusion, oxygen and carbon dioxide travel the same path but in opposite directions.

Diffusion

In diffusion, oxygen and carbon dioxide molecules move between the alveoli and capillaries. The direction of movement is always from an area of greater concentration to one of lesser concentration. In the process, oxygen moves across the alveolar and capillary membranes, dissolves in the plasma, and then passes through the red blood cell (RBC) membrane. Carbon dioxide moves in the opposite direction.

The interesting thing about interstitial spaces

The epithelial membranes lining the alveoli and capillaries must be intact. Both the alveolar epithelium and the capillary endothelium are composed of a single layer of cells. Between these layers are tiny interstitial spaces filled with elastin and collagen. Thickening in the interstitial spaces can slow diffusion.

Move it on over! This is my turf now.

From the RBCs to the alveoli

Normally, oxygen and carbon dioxide move easily through all of these layers. Oxygen moves from the alveoli into the bloodstream, where it's taken up by hemoglobin in the RBCs. When oxygen arrives in the bloodstream, it displaces carbon dioxide (the by-product of metabolism), which diffuses from RBCs into the blood and then to the alveoli.

To bind or not to bind

Most transported oxygen binds with hemoglobin to form oxyhemoglobin; however, a small portion dissolves in the plasma. The portion of oxygen that dissolves in plasma can be measured as the partial pressure of oxygen in arterial blood, or Pao_2.

Now I get it!

What happens in ventilation-perfusion mismatch

Ideally, the amount of air in the alveoli (a reflection of ventilation) matches the amount of blood in the capillaries (a reflection of perfusion). This allows gas exchange to proceed smoothly.

This ventilation-perfusion ($\dot{V}/\dot{Q}$) ratio is actually unequal: The alveoli receive air at a rate of approximately 4 L/minute, while the capillaries supply blood at a rate of about 5 L/minute. This creates a $\dot{V}/\dot{Q}$ mismatch of 4:5, or 0.8.

Normal
In the normal lung, ventilation closely matches perfusion.

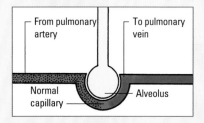

Dead-space ventilation
Normal ventilation without adequate perfusion usually results from a perfusion defect such as pulmonary embolism.

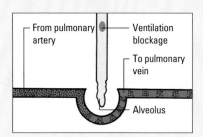

Shunt
Perfusion without adequate ventilation usually results from airway obstruction, particularly that caused by acute diseases, such as atelectasis and pneumonia.

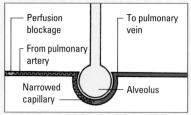

Silent unit
Inadequate ventilation and perfusion usually stems from multiple causes, such as pulmonary embolism with resultant acute respiratory distress syndrome and emphysema.

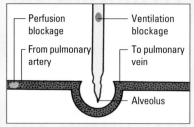

Blood with CO_2	Blood with O_2	Blood with CO_2 and O_2

Now I get it!

Exchanging gases

Gas exchange occurs very rapidly in the millions of tiny, thin-membraned alveoli within the respiratory units. Inside these air sacs, oxygen from inhaled air diffuses into the blood while carbon dioxide diffuses from the blood into the air and is exhaled. Blood then circulates throughout the body, delivering oxygen and picking up carbon dioxide. Finally, the blood returns to the lungs to be oxygenated again.

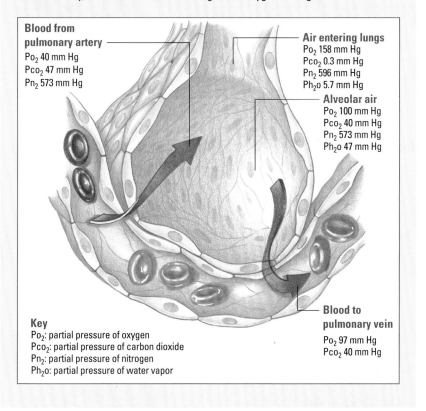

Blood from pulmonary artery
Po_2 40 mm Hg
Pco_2 47 mm Hg
Pn_2 573 mm Hg

Air entering lungs
Po_2 158 mm Hg
Pco_2 0.3 mm Hg
Pn_2 596 mm Hg
Ph_2o 5.7 mm Hg

Alveolar air
Po_2 100 mm Hg
Pco_2 40 mm Hg
Pn_2 573 mm Hg
Ph_2o 47 mm Hg

Blood to pulmonary vein
Po_2 97 mm Hg
Pco_2 40 mm Hg

Key
Po_2: partial pressure of oxygen
Pco_2: partial pressure of carbon dioxide
Pn_2: partial pressure of nitrogen
Ph_2o: partial pressure of water vapor

After oxygen binds to hemoglobin, RBCs travel to the tissues. Through cellular diffusion, internal respiration occurs when RBCs release oxygen and absorb carbon dioxide. The RBCs then transport the carbon dioxide back to the lungs for removal during expiration. (See *Exchanging gases*.)

Acid–base balance

Oxygen taken up in the lungs is transported to the tissues by the circulatory system, which exchanges it for carbon dioxide produced by metabolism in body cells. Because carbon dioxide is more soluble than oxygen, it dissolves in the blood. In the blood, most of the carbon dioxide forms bicarbonate (base); smaller amounts form carbonic acid (acid).

> It takes cooperation! I send the messages…

Respiratory responses

The lungs control bicarbonate levels by converting bicarbonate to carbon dioxide and water for excretion. In response to signals from the medulla, the lungs can change the rate and depth of breathing. This change allows for adjustments in the amount of carbon dioxide lost to help maintain acid-base balance.

Respiratory alkalosis

For example, in respiratory *alkalosis* (a condition resulting from excess bicarbonate retention), the rate and depth of ventilation decrease so that carbon dioxide can be retained; this increases carbonic acid levels.

Respiratory acidosis

In *respiratory acidosis* (a condition resulting from excess acid retention or excess bicarbonate loss), the lungs increase the rate and depth of ventilation to eliminate excess carbon dioxide, thus reducing carbonic acid levels.

> …and we change the rate and depth of breathing. Together we adjust levels of carbon dioxide and maintain acid-base balance.

Imbalance woes

When the lungs don't function properly, an acid-base imbalance results. For example, they can cause respiratory acidosis through *hypoventilation* (reduced rate and depth of alveolar ventilation), which leads to carbon dioxide retention. Conversely, respiratory alkalosis results from *hyperventilation* (increased rate and depth of alveolar ventilation), which leads to carbon dioxide elimination.

Quick quiz

1. Which of the following structures is the chief respiratory unit for gas exchange?
    A. Acinus
    B. Alveoli
    C. Terminal bronchioles
    D. Pulmonary arteries

Answer: A. The acinus is the chief respiratory unit for gas exchange.

2. How many lobes does the right lung have?
    A. Six
    B. Two
    C. Three
    D. One

Answer: C. The right lung has three lobes.

3. During external gas exchange, oxygen and carbon dioxide diffusion occurs in the:
    A. venules.
    B. alveoli.
    C. red blood cells.
    D. body tissues.

Answer: B. Oxygen and carbon dioxide diffusion occurs in the alveoli.

4. When oxygen passes through the alveoli into the bloodstream, it binds with hemoglobin to form:
    A. red blood cells.
    B. carbon dioxide.
    C. nitrogen.
    D. oxyhemoglobin.

Answer: D. When oxygen passes through the alveoli into the bloodstream, it binds with hemoglobin to form oxyhemoglobin.

Scoring

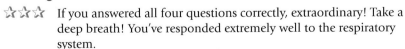

    If you answered all four questions correctly, extraordinary! Take a deep breath! You've responded extremely well to the respiratory system.

    If you answered three questions correctly, fascinating! You're breezing through these systems like a whirlwind.

    If you answered fewer than three questions correctly, get inspired! Perhaps you'll catch your second wind the next time through the chapter.

Just for fun!

Unscramble the words on the left to discover what occurs when the body needs increased oxygenation. Then draw lines linking each process to its appropriate muscular responses.

CODREF RAISINPOINT

_ _ _ _ _ _

_ _ _ _ _ _ _ _ _ _

VATICE PEANIXTRIO

_ _ _ _ _ _

_ _ _ _ _ _ _ _ _ _

A. Pectoral muscles raise the chest to increase the anteroposterior diameter.
B. Internal intercostal muscles contract to shorten the chest's transverse diameter.
C. Scalene muscles elevate, fix, and expand the upper chest.
D. Posterior trapezius muscles raise the thoracic cage.
E. Abdominal rectus muscles pull down the lower chest, depressing the ribs.
F. Sternocleidomastoid muscles raise the sternum.

Answer: FORCED INSPIRATION: A, C, D, F; ACTIVE EXPIRATION: B, E

Selected References

American Lung Association. (2016). *How lungs work.* Retrieved from http://www.lung.org/lung-health-and-diseases/how-lungs-work/

American Thoracic Society. (2016). *Anatomy and function of the normal lung.* Retrieved from https://www.thoracic.org/copd-guidelines/for-patients/anatomy-and-function-of-the-normal-lung.php

Barrow, A., & Pandit, J. (2014). Lung ventilation and the physiology of breathing. *Surgery, 32*(5), 221–227. doi: 10.1016/j.mpsur.2014.02.010. ISSN: 0263-9319.

Canadian Cancer Society. (2016). *Anatomy and physiology of the lung.* Retrieved from http://www.cancer.ca/en/cancer-information/cancer-type/lung/anatomy-and-physiology/?region=on

Forbes, H., & Watt, E. (2015). *Jarvis's physical examination and health assessment* (2nd ed.). Australia: Elsevier.

McCance, K., & Huether, S. (2010). *Pathophysiology: The biologic basis for disease in adults and children* (6th ed.). Maryland Heights, MO: Mosby.

Palmer, B. (2012). Evaluation and treatment of respiratory alkalosis. *American Journal of Kidney Disease, 60*(5), 834–838. doi: 10.1053/j.ajkd.2012.03.025.

Qureshi, S. (2015). Measurement of respiratory function: An update on gas exchange. *Anesthesia & Intensive Care Medicine, 16*(2), 68–73. doi: 10.1016/j.mpaic.2014.11.005.

Chapter 12

Gastrointestinal system

Just the facts

In this chapter, you'll learn:

◆ two major components of the GI system

◆ phases of digestion

◆ functions of GI hormones

◆ sites and mechanisms of gastric secretions.

A look at the GI system

The GI system has two major components: the *alimentary canal* (also called the *GI tract*) and the *accessory GI organs*.

The GI tract serves two major functions:

1. *digestion,* or the breaking down of food and fluid into simple chemicals that can be absorbed into the bloodstream and transported throughout the body
2. *elimination* of waste products through excretion of stool.

Alimentary canal

The alimentary canal is a hollow muscular tube that begins in the mouth and extends to the anus. It includes the pharynx, esophagus, stomach, small intestine, and large intestine. (See *Structures of the GI system*, page 192.) The wall of the alimentary canal is made up of several layers.

What goes in must come out. Digestion and excretion are the GI tract's major functions.

Body shop

Structures of the GI system

The GI system includes the alimentary canal (pharynx, esophagus, stomach, and small and large intestines) and the accessory organs (liver, biliary duct system, and pancreas). These structures are illustrated below.

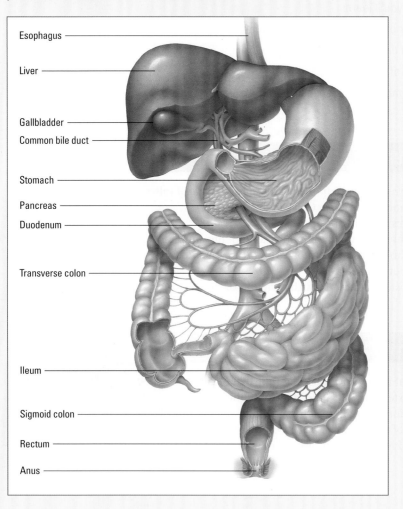

Esophagus

Liver

Gallbladder

Common bile duct

Stomach

Pancreas

Duodenum

Transverse colon

Ileum

Sigmoid colon

Rectum

Anus

Mouth

The mouth (also called the *buccal cavity* or *oral cavity*) is bounded by the lips (labia), cheeks, palate (roof of the mouth), and tongue and contains the teeth. Ducts connect the mouth with the three major pairs of salivary glands (parotid, submandibular, sublingual).

These glands secrete saliva to moisten food during chewing. The mouth initiates the mechanical breakdown of food. (See *Oral cavity*, page 194.)

> Moisten, chew, and break us down.

Pharynx

The *pharynx* is a cavity that extends from the base of the skull to the esophagus. The pharynx aids swallowing by grasping food and propelling it toward the esophagus. When food enters the pharynx, the *epiglottis* (a flap of connective tissue) closes over the trachea to prevent aspiration.

Esophagus

The *esophagus* is a muscular tube that extends from the pharynx through the mediastinum to the stomach. Swallowing triggers the passage of food from the pharynx to the esophagus. The cricopharyngeal sphincter—a sphincter at the upper border of the esophagus—must relax for food to enter the esophagus. Peristalsis propels liquids and solids through the esophagus into the stomach.

Stomach

The *stomach* is a collapsible, pouchlike structure in the left upper portion of the abdominal cavity, just below the diaphragm. Its upper border attaches to the lower end of the esophagus. The lateral surface of the stomach is called the *greater curvature*; the medial surface, the *lesser curvature*.

Size does matter

The size of the stomach varies with the degree of distention. Overeating can cause marked distention, which pushes on the diaphragm and causes shortness of breath.

The Fab Four

The stomach has four main regions:
1. The *cardia* lies near the junction of the stomach and esophagus.

2. The *fundus* is an enlarged portion above and to the left of the esophageal opening into the stomach.
3. The *body* is the middle portion of the stomach.
4. The *pylorus* is the lower portion, lying near the junction of the stomach and duodenum.

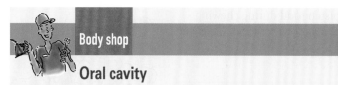

Body shop

Oral cavity

The mouth, or oral cavity, is bounded by the lips (labia), cheeks, palate (roof of the mouth), and tongue. The mouth initiates the mechanical breakdown of food.

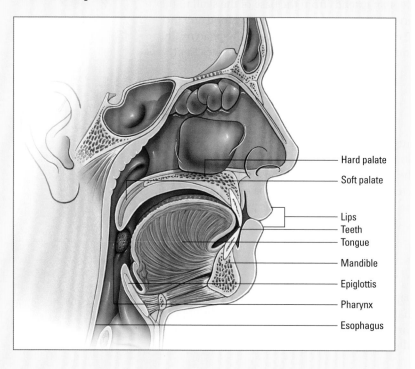

Hard palate

Soft palate

Lips
Teeth
Tongue

Mandible

Epiglottis

Pharynx

Esophagus

Just passing through

The stomach has several functions, including:
- serving as a temporary storage area for food
- beginning digestion
- breaking down food into *chyme*, a semifluid substance
- moving the gastric contents into the small intestine.

The small intestine is a tube that measures about 20' in length. It's the longest organ of the GI tract.

Small intestine

The *small intestine* is a tube that measures about 20' (6 m) in length. It's the longest organ of the GI tract and has three major divisions:
- The *duodenum* is the shortest and most superior division.
- The *jejunum* is the middle portion.
- The *ileum* is the longest and most inferior portion. (See *A look at special cells*, page 196.)

Intestinal wall

The intestinal wall has structural features that significantly increase its absorptive surface area. These features include *plicae circulares*—circular folds of the intestinal mucosa, or mucous membrane lining.

Free villi

Villi and microvilli are also intestinal wall features that increase the absorptive area of the intestinal wall. *Villi* are fingerlike projections on the mucosa. *Microvilli* are tiny cytoplasmic projections on the surface of epithelial cells.

Other structures

The small intestine also contains intestinal crypts, Peyer's patches, and Brunner's glands:
- *Intestinal crypts* are simple glands lodged in the grooves separating villi.
- *Peyer's patches* are collections of lymphatic tissue within the submucosa.
- *Brunner's glands* secrete mucus.

Functions

Functions of the small intestine include:
- completing food digestion
- absorbing food molecules through its wall into the circulatory system, which then delivers them to body cells
- secreting hormones that help control the secretion of bile, pancreatic fluid, and intestinal fluid.

Completing digestion, absorbing food, and controlling the secretion of bile... the small intestine has a lot of work to do!

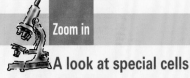

A look at special cells

This cross section of the stomach shows the G cells (which secrete gastrin) in the pyloric glands. The cross section of the duodenum and jejunum shows the S cells (which secrete secretin) in the duodenal and jejunal glands.

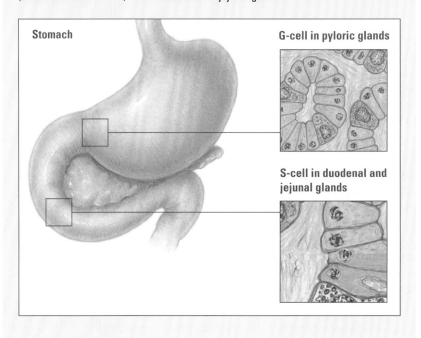

Large intestine

The *large intestine* extends from the ileocecal valve (the valve between the ileum of the small intestine and the first segment of the large intestine) to the anus. It has six segments:

1. The *cecum*, a saclike structure, makes up the first few inches.
2. The *ascending colon* rises on the right posterior abdominal wall and then turns sharply under the liver at the hepatic flexure.
3. The *transverse colon* is situated above the small intestine, passing horizontally across the abdomen and below the liver, stomach, and spleen. At the left colic flexure, also known as the *splenic flexure*, it turns downward.

4. The *descending colon* starts near the spleen and extends down the left side of the abdomen into the pelvic cavity.
5. The *sigmoid colon* descends through the pelvic cavity, where it becomes the rectum.
6. The *rectum*, the last few inches of the large intestine, terminates at the *anus*, which is the external opening of the large intestine that allows expulsion of waste products.

Functions

The functions of the large intestine include absorbing water, secreting mucus, and eliminating digestive wastes.

GI tract wall structures

The wall of the GI tract consists of several layers. These layers are the mucosa, submucosa, tunica muscularis, and visceral peritoneum.

Mucosa

The *mucosa*, the innermost layer, also called the *tunica mucosa*, consists of epithelial and surface cells and loose connective tissue. *Villi*, finger-like projections of the mucosa, secrete gastric and protective juices and absorb nutrients.

Submucosa

The *submucosa*, also called the *tunica submucosa*, encircles the mucosa. It's composed of loose connective tissue, blood and lymphatic vessels, and a nerve network called the *submucosal plexus*, or *Meissner's plexus*.

Tunica muscularis

The *tunica muscularis*, which lies around the submucosa, is composed of skeletal muscle in the mouth, pharynx, and upper esophagus.

Fibers, fibers everywhere

Elsewhere in the tract, the tunica muscularis is made up of longitudinal and circular smooth muscle fibers. During peristalsis, longitudinal fibers shorten the lumen length and circular fibers reduce the lumen diameter. At points along the tract, circular fibers thicken to form sphincters.

Pucker pouches

In the large intestine, these fibers gather into three narrow bands (*taeniae coli*) down the middle of the colon and pucker the intestine into characteristic pouches (*haustra*).

Networking is key

Between the two muscle layers lies another nerve network—the *myenteric plexus*, also known as *Auerbach's plexus*. The stomach wall contains a third muscle layer made up of oblique fibers. (See *Features of the GI tract wall*.)

Visceral peritoneum

The *visceral peritoneum* is the GI tract's outer covering. It covers most of the abdominal organs and lies next to an identical layer, the *parietal peritoneum*, which lines the abdominal cavity.

A double-layered fold

The visceral peritoneum becomes a double-layered fold around the blood vessels, nerves, and lymphatics. It attaches the jejunum and ileum to the posterior abdominal wall to prevent twisting. A similar fold attaches the transverse colon to the posterior abdominal wall.

A visceral peritoneum by any other name

The visceral peritoneum has many names. In the esophagus and rectum, it's called the *tunica adventitia*; elsewhere in the GI tract, it's called the *tunica serosa*.

So many ways of saying one thing! The visceral peritoneum is known as the *tunica adventitia* in one part of the body and the *tunica serosa* in another.

GI tract innervation

Distention of the submucosal plexus stimulates transmission of nerve signals to the smooth muscle, which initiates peristalsis and mixing contractions.

Parasympathetic stimulation

Parasympathetic stimulation of the vagus nerve (for most of the intestines) and the sacral spinal nerves (for the descending colon and rectum) increases gut and sphincter tone. It also increases the frequency, strength, and velocity of smooth-muscle contractions as well as motor and secretory activities.

Zoom in

Features of the GI tract wall

Several layers—the tunica mucosa, tunica submucosa, and tunica muscularis—form the wall of the GI tract. This illustration depicts the cellular anatomy of the wall, including special features, such as the villi, the peritoneum, the muscles, and a nerve network.

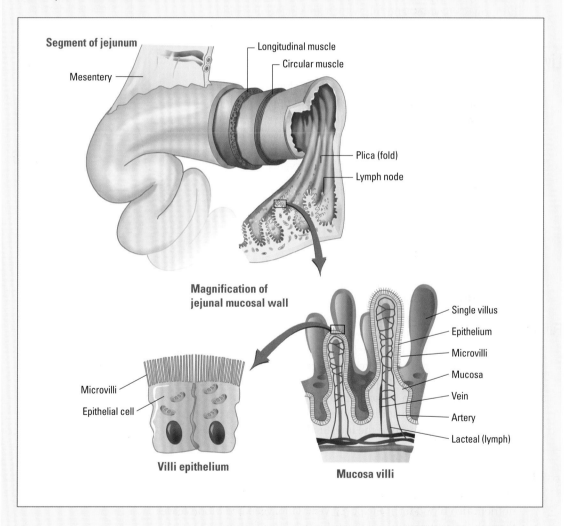

Segment of jejunum

Mesentery

Longitudinal muscle

Circular muscle

Plica (fold)

Lymph node

Magnification of jejunal mucosal wall

Single villus

Epithelium

Microvilli

Mucosa

Vein

Artery

Lacteal (lymph)

Microvilli

Epithelial cell

Villi epithelium

Mucosa villi

Sympathetic stimulation

Sympathetic stimulation, by way of the spinal nerves from levels T6 to L2, reduces peristalsis and inhibits GI activity.

Accessory GI organs

Accessory GI organs—the liver, gallbladder, and pancreas—contribute hormones, enzymes, and bile, which are vital to digestion.

A little respect please! I am the largest gland in the body.

Liver

The body's largest gland, the 3-lb (1.4-kg), highly vascular liver is enclosed in a fibrous capsule in the right upper quadrant of the abdomen. The *lesser omentum*, a fold of peritoneum, covers most of the liver and anchors it to the lesser curvature of the stomach. The *hepatic artery* and *hepatic portal vein*, as well as the common bile duct and hepatic veins, pass through the lesser omentum.

Lobes and lobules

The liver consists of four lobes:
- left lobe
- right lobe
- caudate lobe (behind the right lobe)
- quadrate lobe (behind the left lobe).

Function…

The liver's functional unit, the *lobule*, consists of a plate of hepatic cells, or *hepatocytes*, that encircle a central vein and radiate outward. Separating the hepatocyte plates from each other are *sinusoids*, the liver's capillary system. Reticuloendothelial macrophages (Kupffer cells) that line the sinusoids remove bacteria and toxins that have entered the blood through the intestinal capillaries. (See *A look at a liver lobule.*)

…and flow

The sinusoids carry oxygenated blood from the hepatic artery and nutrient-rich blood from the portal vein. Unoxygenated blood leaves through the central vein and flows through hepatic veins to the inferior vena cava.

Zoom in

A look at a liver lobule

The liver's functional unit is called a *lobule*. It consists of a plate of hepatic cells, or hepatocytes, that encircle a central vein and radiate outward. Separating the hepatocyte plates from each other are *sinusoids*, which serve as the liver's capillary system. Sinusoids carry oxygenated blood from the hepatic artery and nutrient-rich blood from the portal vein.

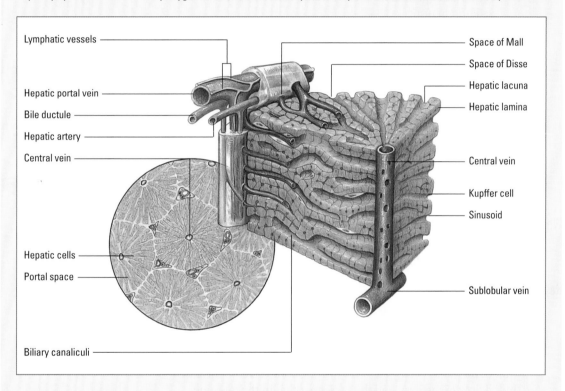

Lymphatic vessels

Hepatic portal vein

Bile ductule

Hepatic artery

Central vein

Hepatic cells

Portal space

Biliary canaliculi

Space of Mall

Space of Disse

Hepatic lacuna

Hepatic lamina

Central vein

Kupffer cell

Sinusoid

Sublobular vein

Ducts

The ducts can be thought of as a subway system that transports bile through the GI tract. *Bile* is a greenish liquid composed of water, cholesterol, bile salts, and phospholipids. It exits through bile ducts (canaliculi) that merge into the right and left hepatic ducts to form the common hepatic duct. This duct joins the cystic duct from the gallbladder to form the common bile duct that leads to the duodenum.

Job description

The liver serves several important functions:
- It plays an important role in carbohydrate metabolism.
- It detoxifies various endogenous and exogenous toxins in plasma.
- It synthesizes plasma proteins, nonessential amino acids, and vitamin A.
- It stores essential nutrients, such as vitamins K, D, and B_{12} and iron.
- It removes ammonia from body fluids, converting it to urea for excretion in urine.
- It helps regulate blood glucose levels by storing and releasing glucose as needed.
- It secretes bile.
- It processes hemoglobin for utilization of its iron content.
- It regulates blood clotting.

My role in carbohydrate metabolism is very important. I also detoxify toxins in plasma.

Function of bile

Bile has several functions, including:
- emulsifying (breaking down) fat
- promoting intestinal absorption of fatty acids, cholesterol, and other lipids.

When bile salts are MIA

When bile salts are absent from the intestinal tract, lipids are excreted and fat-soluble vitamins are poorly absorbed.

Report on bile production

The liver recycles about 80% of bile salts into bile, combining them with bile pigments (biliverdin and bilirubin, the waste products of red blood cell breakdown) and cholesterol. The liver continuously secretes this alkaline bile. Bile production may increase from stimulation of the vagus nerve, release of the hormone secretin, increased blood flow in the liver, and the presence of fat in the intestine. (See *GI hormones: Production and function*.)

Gallbladder

The *gallbladder* is a pear-shaped organ joined to the ventral surface of the liver by the cystic duct. It's covered with visceral peritoneum.

On the job

The gallbladder stores and concentrates bile produced by the liver. It also releases bile into the common bile duct (formed by the cystic duct and common hepatic duct) for delivery to the duodenum in response to the contraction and relaxation of the sphincter of Oddi.

Now I get it!

GI hormones: Production and function

When stimulated, GI structures secrete four hormones. Each hormone plays a different role in digestion.

Hormone and production site	Stimulating factor or agent	Function
Gastrin Produced in pyloric antrum and duodenal mucosa	• Pyloric antrum distention • Vagal stimulation • Protein digestion products • Alcohol	Stimulates gastric secretion and motility
Gastric inhibitory peptides Produced in duodenal and jejunal mucosa	• Gastric acid • Fats • Fat digestion products	Inhibits gastric secretion and motility
Secretin Produced in duodenal and jejunal mucosa	• Gastric acid • Fat digestion products • Protein digestion products	Stimulates secretion of bile and alkaline pancreatic fluid
Cholecystokinin Produced in duodenal and j ejunal mucosa	• Fat digestion products • Protein digestion products	Stimulates gallbladder contraction and secretion of enzyme-rich pancreatic fluid

Pancreas

The *pancreas* is a somewhat flat organ that lies behind the stomach. Its head and neck extend into the curve of the duodenum, and its tail lies against the spleen. (See *A look at the biliary tract,* page 204.) It performs both exocrine and endocrine functions.

Exocrine function

The pancreas's exocrine function involves scattered cells that secrete more than 1,000 mL of digestive enzymes every day. Lobules and lobes of the clusters (*acini*) of enzyme-producing cells release their secretions into ducts that merge into the pancreatic duct.

Memory jogger

To remember the difference between exocrine and endocrine, keep in mind that **ex**ocrine refers to **ex**ternal, and **endo**crine refers to **in**ternal.

A look at the biliary tract

Together, the gallbladder and pancreas constitute the biliary tract. The structures of the biliary tract are depicted in the illustration below.

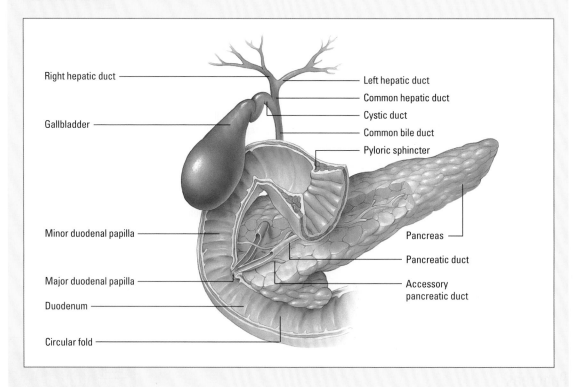

Endocrine function

The endocrine function of the pancreas is performed by the islets of Langerhans, which are located between the acinar cells.

Alpha, beta…but no omega

There are two types of islet cells: alpha and beta. More than 1 million islets house these two cell types. Beta cells secrete *insulin* to promote carbohydrate metabolism; alpha cells secrete *glucagon*, a hormone that stimulates glycogenolysis in the liver. Both hormones flow directly into the blood. Their release is stimulated by blood glucose levels.

Pancreatic duct

Running the length of the pancreas, the *pancreatic duct* joins the bile duct from the gallbladder before entering the duodenum. Vagal stimulation and release of the hormones secretin and cholecystokinin control the rate and amount of pancreatic secretion.

Digestion and elimination

Digestion starts in the oral cavity, where chewing (*mastication*), salivation (the beginning of starch digestion), and swallowing (*deglutition*) take place.

When a person swallows, the hypopharyngeal sphincter in the upper esophagus relaxes, allowing food to enter the esophagus. (See *What happens in swallowing*, page 206.)

Chewing, salivation, and swallowing. Let digestion begin!

Long day's journey into the stomach

In the esophagus, the glossopharyngeal nerve activates peristalsis, which moves the food down toward the stomach. As food passes through the esophagus, glands in the esophageal mucosal layer secrete mucus, which lubricates the bolus and protects the mucosal membrane from damage caused by poorly chewed foods.

Cephalic phase of digestion

By the time the food bolus is traveling toward the stomach, the *cephalic phase* of digestion has already begun. In this phase, the stomach secretes digestive juices (hydrochloric acid [HCl] and pepsin).

Gastric phase of digestion

When food enters the stomach through the cardiac sphincter, the stomach wall stretches, initiating the gastric phase of digestion. In this phase, distention of the stomach wall stimulates the stomach to release *gastrin*.

Gastrin

Gastrin stimulates the stomach's motor functions and secretion of gastric juice by the gastric glands. Highly acidic (pH of 0.9 to 1.5), these digestive secretions consist mainly of pepsin, HCl, intrinsic factor, and proteolytic enzymes. (See *Sites and mechanisms of gastric secretion*, page 207.)

What happens in swallowing

Before peristalsis can begin, the neural pattern that initiates swallowing must occur. This process is described here and illustrated below:

• Food pushed to the back of the mouth stimulates swallowing receptor areas that surround the pharyngeal opening.

• These receptor areas transmit impulses to the brain by way of the sensory portions of the trigeminal and glosso-pharyngeal nerves.

• The brain's swallowing center then relays motor impulses to the esophagus by way of the trigeminal, glossopharyngeal, vagus, and hypoglossal nerves, causing swallowing to occur.

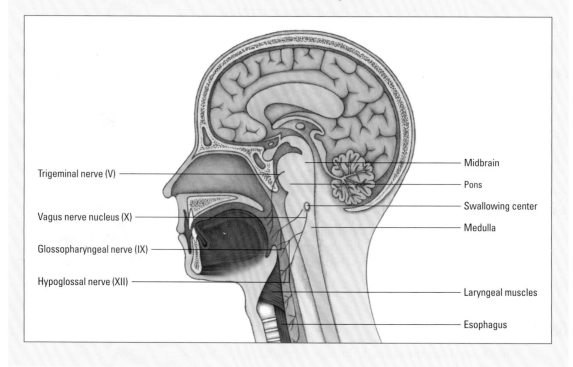

Trigeminal nerve (V)

Vagus nerve nucleus (X)

Glossopharyngeal nerve (IX)

Hypoglossal nerve (XII)

Midbrain

Pons

Swallowing center

Medulla

Laryngeal muscles

Esophagus

Intestinal phase of digestion

Normally, except for alcohol, minimal food absorption occurs in the stomach. Peristaltic contractions churn the food into tiny particles and mix it with gastric juices, forming *chyme*. Next, stronger peristaltic waves move the chyme into the antrum, where it backs up against the pyloric sphincter before being released into the duodenum, triggering the intestinal phase of digestion.

Now I get it!

Sites and mechanisms of gastric secretion

The body of the stomach lies between the lower esophageal, or cardiac, sphincter and the pyloric sphincter. Between these sphincters lie the fundus, body, antrum, and pylorus. These areas have a rich variety of mucosal cells that help the stomach carry out its tasks.

Glands and gastric secretions

Cardiac glands, pyloric glands, and gastric glands secrete 2 to 3 L of gastric juice daily through the stomach's gastric pits. Here are the details:
- Both the *cardiac gland* (near the lower esophageal sphincter [LES]) and the *pyloric gland* (near the pylorus) secrete thin mucus.
- The *gastric gland* (in the body and fundus) secretes hydrochloric acid (HCl), pepsinogen, intrinsic factor, and mucus.

Protection from self-digestion

Specialized cells line the gastric glands, gastric pits, and surface epithelium. Mucous cells in the necks of the gastric glands produce thin mucus. Mucous cells in the surface epithelium produce an alkaline mucus. Both substances lubricate food and protect the stomach from self-digestion by corrosive enzymes.

Other secretions

Argentaffin cells produce gastrin, which stimulates gastric secretion and motility. *Chief cells* produce pepsinogen, which breaks proteins down into polypeptides. Large parietal cells scattered throughout the fundus secrete HCl and intrinsic factor. HCl degrades pepsinogen, maintains acid environment, and inhibits excess bacteria growth. Intrinsic factor promotes vitamin B_{12} absorption in the small intestine.

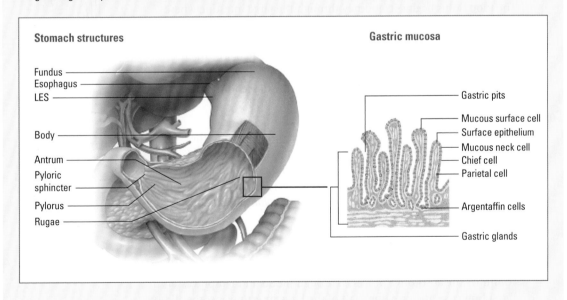

Stomach structures

Fundus
Esophagus
LES
Body
Antrum
Pyloric sphincter
Pylorus
Rugae

Gastric mucosa

Gastric pits
Mucous surface cell
Surface epithelium
Mucous neck cell
Chief cell
Parietal cell
Argentaffin cells
Gastric glands

Stomach emptying

The rate of stomach emptying depends on several factors, including gastrin release, neural signals generated when the stomach wall distends, and the *enterogastric reflex*. In this reaction, the duodenum releases secretin and gastric-inhibiting peptide, and the jejunum secretes cholecystokinin—all of which act to decrease gastric motility.

Small intestine

The small intestine performs most of the work of digestion and absorption. (See *Small intestine: How form affects absorption*.)

Small but mighty

In the small intestine, intestinal contractions and various digestive secretions break down carbohydrates, proteins, and fats—actions that enable the intestinal mucosa to absorb these nutrients into the bloodstream (along with water and electrolytes). These nutrients are then available for use by the body.

By the time chyme passes through the small intestine and enters the ascending colon of the large intestine, it has been reduced to mostly indigestible substances.

Large intestine

The food bolus begins its journey through the large intestine where the ileum and cecum join with the ileocecal pouch. Then the bolus moves up the ascending colon and past the right abdominal cavity to the liver's lower border. It crosses horizontally below the liver and stomach, by way of the transverse colon, and descends the left abdominal cavity to the iliac fossa through the descending colon.

Continuing journey of the food bolus

From there, the bolus travels through the sigmoid colon to the lower midline of the abdominal cavity, then to the rectum, and finally to the anal canal. The anus opens to the exterior through two sphincters. The *internal sphincter* contains thick, circular smooth muscle under autonomic control; the *external sphincter* contains skeletal muscle under voluntary control.

Role in absorption

The large intestine produces no hormones or digestive enzymes; it continues the absorptive process. Through blood and lymph vessels in the submucosa, the proximal half of the large intestine absorbs all but about 100 mL of the remaining water in the colon each day. It also absorbs large amounts of sodium and chloride.

It only takes a whiff! Digestive juices are secreted in response to smelling, tasting, chewing, or thinking about food.

Now I get it!

Small intestine: How form affects absorption

Nearly all digestion and absorption takes place in the 20' (6 m) of small intestine. The structure of the small intestine, as shown below, is key to digestion and absorption.

Specialized mucosa

Multiple projections of the intestinal mucosa increase the surface area for absorption several hundredfold, as shown in the enlarged view at bottom left.

Circular folds are covered by villi. Each villus contains a lymphatic vessel (*lacteal*), a venule, capillaries, an arteriole, nerve fibers, and smooth muscle.

Each villus is densely fringed with about 2,000 microvilli making it resemble a fine brush. The villi are lined with columnar epithelial cells, which dip into the lamina propria between the villi to form intestinal glands (*crypts of Lieberkühn*).

Types of epithelial cells

The type of epithelial cell dictates its function. Mucus-secreting goblet cells are found on and between the villi on the crypt mucosa. In the proximal duodenum, specialized Brunner's glands also secrete large amounts of mucus to lubricate and protect the duodenum from potentially corrosive acidic chyme and gastric juices.

Duodenal *argentaffin cells* produce the hormones secretin and cholecystokinin. *Undifferentiated cells* deep within the intestinal glands replace the epithelium. *Absorptive cells* consist of many tightly packed microvilli over a plasma membrane that contains transport mechanisms for absorption and produces enzymes for the final step in digestion.

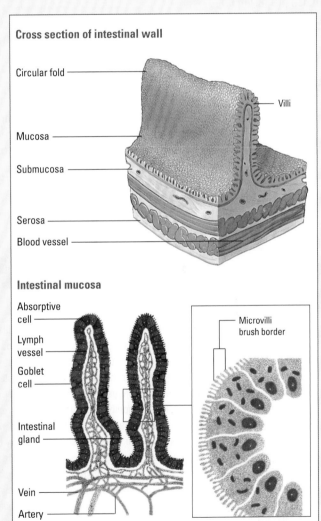

Cross section of intestinal wall

Circular fold

Villi

Mucosa

Submucosa

Serosa

Blood vessel

Intestinal mucosa

Absorptive cell

Lymph vessel

Goblet cell

Intestinal gland

Vein

Artery

Microvilli brush border

Intestinal glands

The intestinal glands primarily secrete a watery fluid that bathes the villi with chyme particles. Fluid production results from local irritation of nerve cells and, possibly, from hormonal stimulation by secretin and cholecystokinin. The microvillous brush border secretes various hormones and digestive enzymes that catalyze final nutrient breakdown.

Bacterial action

The large intestine harbors the bacteria *Escherichia coli*, *Enterobacter aerogenes*, *Clostridium perfringens*, and *Lactobacillus bifidus*. All of these bacteria help synthesize vitamin K and break down cellulose into a usable carbohydrate. Bacterial action also produces *flatus*, which helps propel stool toward the rectum.

Mucosa preparing for...

In addition, the mucosa of the large intestine produces *alkaline secretions* from tubular glands composed of goblet cells. This alkaline mucus lubricates the intestinal walls as food pushes through, protecting the mucosa from acidic bacterial action.

...mass movement

In the lower colon, long and relatively sluggish contractions cause propulsive waves, or *mass movements*. Normally occurring several times per day, these movements propel intestinal contents into the rectum and produce the urge to defecate.

Defecation normally results from the *defecation reflex*, a sensory and parasympathetic nerve-mediated response, along with the voluntary relaxation of the external anal sphincter. (See *GI changes with aging*.)

We're not all bad. Some of us bacteria aid GI function.

Senior moment

GI changes with aging

The physiologic changes that accompany aging usually prove less debilitating in the GI system than in most other body systems. This is partly due to the enhanced functional reserve capacity of the GI tract. However, like other body systems that undergo aging changes, there is a decrease in the production of new cells in the GI system, and tissues become more at risk for damage. Normal changes include diminished mucosal elasticity and reduced GI secretions, which, in turn, modify some processes—for example, digestion and absorption. GI tract motility, bowel wall and anal sphincter tone, and abdominal muscle strength also may decrease with age. Any of these changes may cause complaints in an older patient, ranging from loss of appetite to constipation. In addition, the incidence of GI cancer rises with aging, especially in the stomach and colon.

Changes in the oral cavity also occur. Tooth enamel wears away, leaving the teeth prone to cavities. Periodontal disease increases and the number of taste buds declines. The sense of smell diminishes and salivary gland secretion decreases, leading to appetite loss.

Liver changes

Normal physiologic changes in the liver include decreased liver weight, reduced regenerative capacity, and decreased blood flow to the liver. Because hepatic enzymes involved in oxidation and reduction markedly decline with age, the liver metabolizes drugs and detoxifies substances less efficiently.

Quick quiz

1. Which component of the GI system completes food digestion?
    A. Stomach
    B. Gallbladder
    C. Small intestine
    D. Large intestine

Answer: C. The small intestine completes the process of digestion.

2. One of the functions of the liver is:
    A. regulating gastrin secretion.
    B. storing vitamins B_6, C, and E.
    C. storing plasma proteins and carbohydrates.
    D. detoxifying endogenous and exogenous toxins in plasma.

Answer: D. Among many other functions, the liver is responsible for detoxifying various exogenous and endogenous toxins in plasma.

3. Which GI hormone stimulates gastric secretion and motility?
    A. Gastrin
    B. Gastric inhibitory peptides
    C. Secretin
    D. Cholecystokinin

Answer: A. Gastrin is produced in the pyloric antrum and duodenal end mucosa and stimulates gastric secretion and motility.

4. In which phase of digestion does the stomach secrete the digestive juices hydrochloric acid and pepsin?
    A. Cephalic
    B. Gastric
    C. Intestinal
    D. Stomach emptying

Answer: A. By the time food is traveling toward the stomach, the cephalic phase—during which the stomach secretes digestive juices—has begun.

Scoring

☆☆☆   If you answered all four questions correctly, bravo! You've passed through the GI system with the greatest of ease!

☆☆   If you answered three questions correctly, super! You've chewed the fat of this system and it's time to move on.

☆   If you answered fewer than three questions correctly, don't worry! It might take a little longer to digest this material, but keep at it!

Just for fun!

Unscramble the following words to discover the names of the layers of the GI tract wall. Then draw lines to connect each layer with its particular characteristics.

CAMSOU

_ _ _ _ _ _

SCUBASUMO

_ _ _ _ _ _ _ _ _

ACTINU ACURLISMUS

_ _ _ _ _ _   _ _ _ _ _ _ _ _ _ _

A. The innermost layer
B. Lies around the submucosa
C. Also called the *tunica submucosa*
D. Also called the *tunica mucosa*
E. Consists of epithelial and surface cells and loose connective tissue
F. Composed of loose connective tissue, blood and lymphatic vessels, and a nerve network
G. Contains fingerlike projections called villi
H. Encircles the mucosa
I. Composed of skeletal muscle in the mouth, pharynx, and upper esophagus; elsewhere, is made up of longitudinal and circular smooth muscle fibers

Answer: MUCOSA: A, D, E, G; SUBMUCOSA: C, F, H; TUNICA MUSCULARIS: B, I

Selected Reference

McCance, K. L., & Huether, S. E. (2015). *Pathophysiology: The biologic basis for disease in adults and children*. Retrieved from https://books.google.com/books. ISBN No.: 0323293751

Chapter 13

Nutrition and metabolism

Just the facts

In this chapter, you'll learn:

◆ the roles of carbohydrates, proteins, and lipids in nutrition

◆ the functions of vitamins and minerals in the body

◆ the way in which glucose is turned into energy

◆ the role of hormones in metabolism.

Nutrition

Nutrition refers to the intake, assimilation, and utilization of nutrients. The crucial nutrients in foods must be broken down into components for use by the body. Within cells, the products of digestion undergo further chemical reactions.

Metabolism refers to the sum of these chemical reactions. Through metabolism, food substances are transformed into energy or materials that the body can use or store.

Metabolism involves two processes:

1. *anabolism*—synthesis of simple substances into complex ones
2. *catabolism*—breakdown of complex substances into simpler ones or into energy.

Needed for nutrition

The body needs a continual supply of water and various nutrients for growth and repair. Virtually all nutrients come from digested food. The three major types of nutrients required by the body are carbohydrates, proteins, and lipids.

Needed for metabolism

Vitamins are essential for normal metabolism. They contribute to the enzyme reactions that promote the metabolism of carbohydrates, proteins, and lipids. *Minerals* are also important.

A Food +
B Metabolism =
C Energy

They participate in such essential functions as enzyme metabolism and membrane transfer of essential elements.

Carbohydrates

Carbohydrates are organic compounds composed of carbon, hydrogen, and oxygen that convert to glucose in the body; they yield 4 kcal/g when used for energy.

Simple? Or complex?

Carbohydrates are categorized as simple or complex. Simple carbohydrates include the sugars in fruits, vegetables, dairy products, and foods made with processed sugar. They raise the blood glucose level quickly. Complex carbohydrates include the starches and fiber found in breads, grains, and beans. They raise the blood glucose level more slowly than do simple carbohydrates.

> The energy in nutrients is measured in kilocalories—commonly just called calories. Adults need between 1,600 and 2,800 kcal daily, depending on their age, height, weight, and physical activity.

Let's go for "aride"

Sugars are classified as *monosaccharides, disaccharides,* and *polysaccharides.* Sugars are carbohydrates and function as the body's primary energy source.

Monosaccharides

Monosaccharides are simple sugars that can't be split into smaller units by hydrolysis. They're subdivided into polyhydroxy aldehydes or ketones, based on whether the molecule consists of an aldehyde group or a ketone group.

The OH link

An aldehyde contains the characteristic group CHO. The term *polyhydroxy* refers to the linking of carbon atoms to a hydroxyl (OH) group.

CO is the key to ketone

A *ketone*, on the other hand, contains the carbonyl group CO and carbon groups attached to the carbonyl carbon.

Disaccharides

Disaccharides are synthesized from monosaccharides. A disaccharide molecule consists of two monosaccharides minus a water molecule. Examples of disaccharides include:

- *sucrose*, common table sugar, which is also found in some fruits and vegetables—a combination of a glucose molecule and a fructose molecule
- *lactose*, the sugar found in milk—a combination of a glucose molecule and a galactose molecule
- *maltose*, a sugar used in brewing and distilling—a combination of two glucose molecules.

Polysaccharides

Like disaccharides, *polysaccharides* are synthesized from monosaccharides. A polysaccharide consists of a long chain (*polymer*) of more than 10 monosaccharides linked by glycoside bonds. Polysaccharides are ingested and broken down into simple sugars and then used for fuel. Glycogen is an example of a polysaccharide. The body builds glycogen by using excess sugar (monosaccharides) and stores it for future use. When glycogen reserves are full, the liver converts the excess to fat.

Fiber is another example of a polysaccharide, but it can't be broken down into simple sugars. Thus, the body can't derive energy (fuel) from fiber.

Proteins

Proteins are complex nitrogenous organic compounds containing amino acid chains; some also contain sulfur and phosphorus. Proteins are used mainly for growth and repair of body tissues; when used for energy, they yield 4 kcal/g. Some proteins combine with lipids to form *lipoproteins* or with carbohydrates to form *glycoproteins*.

Amino acids are the building blocks of proteins.

Amino acids

Amino acids are the building blocks of proteins. Each amino acid contains a carbon atom to which a carboxyl (COOH) group and an amino group are attached.

The bonds that link...

Amino acids unite by condensation of the COOH group on one amino acid with the amino group of the adjacent amino acid. This reaction releases a water molecule and creates a linkage called a *peptide bond*.

...and the acids that attract

The sequence and types of amino acids in the chain determine the nature of the protein. Each protein is synthesized on a ribosome as a straight chain. Chemical attractions between the amino acids in various parts of the chain cause the chain to coil or twist into a specific shape. A protein's shape, in turn, determines its function.

Lipids

Lipids are organic compounds that don't dissolve in water but do dissolve in alcohol and other organic solvents. Lipids are a concentrated form of fuel and yield approximately 9 kcal/g when used for energy. The major lipids include fats (the most common lipids), phospholipids, and steroids.

Fats

A fat, or *triglyceride*, contains three molecules of fatty acid combined with one molecule of glycerol. A fatty acid is composed of a chain of carbon atoms with hydrogen and a few oxygen atoms attached. Fatty acid chains vary in length. Long-chain fatty acids (12 or more carbon atoms) are found in most food fats.

The glycerol example

Glycerol, for example, is a three-carbon compound (alcohol) with an OH group attached to each carbon atom. The COOH group on each fatty acid molecule joins to one OH group on the glycerol molecule; this results in the release of a water molecule. Linking of the COOH and OH groups produces an ester linkage.

Phospholipids

Phospholipids are complex lipids that are similar to fat but have a phosphorus- and nitrogen-containing compound that replaces one of the fatty acid molecules. Phospholipids are major structural components of cell membranes.

Steroids

Steroids are complex molecules in which the carbon atoms form four cyclic structures attached to various side chains. They contain no glycerol or fatty acid molecules. Examples of steroids include cholesterol, bile salts, and sex hormones.

Vitamins and minerals

Vitamins are organic compounds that are needed in small quantities for normal metabolism, growth, and development. Vitamins are classified as *water soluble* or *fat soluble*.

Daily routine

Water-soluble vitamins aren't stored in the body and must be replaced daily. Water-soluble vitamins include the B complex and C vitamins.

Stored but not forgotten

Fat-soluble vitamins are dissolved in fat before they are absorbed by the bloodstream. Excess fat-soluble vitamins are stored in the liver and body tissues; therefore, they don't need to be ingested daily. The fat-soluble vitamins include A, D, E, and K. (See *Guide to vitamins and minerals*, pages 218 to 220.)

Eat up! The body can't manufacture enough of us on its own.

Minerals

- Minerals are inorganic substances that play important roles in:
- enzyme metabolism
- membrane transfer of essential compounds
- regulation of acid-base balance
- osmotic pressure
- muscle contractility
- nerve impulse transmission
- growth.

Minerals are found in bones, hemoglobin, thyroxine, teeth, and organs. They're classified as *major minerals* (more than 0.005% of body weight) or *trace minerals* (less than 0.005% of body weight). Major minerals include calcium, chloride, magnesium, phosphorus, potassium, and sodium. Trace minerals include chromium, cobalt, copper, fluorine, iodine, iron, manganese, molybdenum, selenium, and zinc.

(Text continues on page 220)

Guide to vitamins and minerals

Good health requires intake of adequate amounts of vitamins and minerals to meet the body's metabolic needs. A vitamin or mineral excess or deficiency can lead to various disorders. This chart reviews major functions of vitamins and minerals and their food sources.

Vitamin or mineral	Major functions	Food sources
Water-soluble vitamins		
Vitamin C (ascorbic acid)	• Collagen production, fine bone and tooth formation, iodine conservation, healing, red blood cell (RBC) formation, infection resistance	• Fresh fruits and vegetables
Vitamin B_1 (thiamine)	• Blood formation, carbohydrate metabolism, circulation, digestion, growth, learning ability, muscle tone maintenance, central nervous system (CNS) maintenance	• Meats, fish, poultry, pork, molasses, brewer's yeast, brown rice, nuts, wheat germ, whole and enriched grains
Vitamin B_2 (riboflavin)	• RBC formation; energy metabolism; cell respiration; epithelial, eye, and mucosal tissue maintenance	• Meats, fish, poultry, milk, molasses, brewer's yeast, eggs, fruit, green leafy vegetables, nuts, whole grains
Vitamin B_6 (pyridoxine)	• Antibody formation, digestion, deoxyribonucleic acid (DNA) and ribonucleic acid (RNA) synthesis, fat and protein utilization, amino acid metabolism, hemoglobin production, CNS maintenance	• Meats, poultry, bananas, molasses, brewer's yeast, desiccated liver, fish, green leafy vegetables, peanuts, raisins, walnuts, wheat germ, whole grains
Folic acid (folacin, pteroylglutamic acid)	• Cell growth and reproduction, digestion, liver function, DNA and RNA formation, protein metabolism, RBC formation	• Citrus fruits, eggs, green leafy vegetables, milk products, organ meats, seafood, whole grains
Niacin (nicotinic acid, nicotinamide, niacinamide)	• Circulation, cholesterol level reduction, growth, hydrochloric acid production, metabolism (carbohydrate, protein, fat), sex hormone production	• Eggs, lean meats, milk products, organ meats, peanuts, poultry, seafood, whole grains
Vitamin B_{12} (cyanocobalamin)	• RBC formation, cellular and nutrient metabolism, tissue growth, nerve cell maintenance, appetite stimulation	• Beef, eggs, fish, milk products, organ meats, pork

Guide to vitamins and minerals *(continued)*

Vitamin or mineral	Major functions	Food sources
Fat-soluble vitamins		
Vitamin A	• Body tissue repair and maintenance, infection resistance, bone growth, nervous system development, cell membrane metabolism and structure, night vision	• Meat, milk, eggs, butter • Leafy green and yellow vegetables, yellow fruits (sources of carotene—a precursor to vitamin A)
Vitamin D (calciferol)	• Calcium and phosphorus metabolism (bone formation), myocardial function, nervous system maintenance, normal blood clotting	• Egg yolks, organ meats, butter, cod liver oil, fatty fish
Vitamin E (tocopherol)	• Aging retardation, anticlotting factor, diuresis, fertility, lung protection (antipollution), male potency, muscle and nerve cell membrane maintenance, myocardial perfusion, serum cholesterol reduction	• Butter, dark green vegetables, eggs, fruits, nuts, organ meats, vegetable oils, wheat germ
Vitamin K (menadione)	• Liver synthesis of prothrombin and other blood-clotting factors	• Green leafy vegetables, safflower oil, yogurt, liver, molasses • Also manufactured by bacteria that line the GI tract
Minerals		
Calcium	• Blood clotting, bone and tooth formation, cardiac rhythm, cell membrane permeability, muscle growth and contraction, nerve impulse transmission	• Cheese, milk, molasses, yogurt, whole grains, nuts, legumes, leafy vegetables
Chloride	• Maintenance of fluid, electrolyte, acid-base, and osmotic pressure balance	• Fruits, vegetables, table salt
Magnesium	• Acid-base balance, metabolism, protein synthesis, muscle relaxation, cellular respiration, nerve impulse transmission	• Green leafy vegetables, nuts, seafood, cocoa, whole grains
Phosphorus	• Bone and tooth formation, cell growth and repair, energy production	• Eggs, fish, grains, meats, poultry, milk, milk products

(continued)

Guide to vitamins and minerals *(continued)*

Vitamin or mineral	Major functions	Food sources
Potassium	• Heartbeat, muscle contraction, nerve impulse transmission, rapid growth, fluid distribution and osmotic pressure balance, acid-base balance	• Seafood, molasses, vegetables, fruits, nuts
Sodium	• Cellular fluid level maintenance, muscle contraction, acid-base balance, cell permeability, muscle function, nerve impulse transmission	• Cheese, milk, salt, processed foods, canned soups, fast food, soy sauce
Fluoride (fluorine)	• Bone and tooth formation	• Drinking water, seafood
Iodine	• Thyroid hormone production, energy production, metabolism, physical and mental development	• Kelp, salt (iodized), seafood
Iron	• Growth (in children), hemoglobin production, stress and disease resistance, cellular respiration, oxygen transport	• Egg yolks, meats, poultry, wheat germ, liver, oysters, enriched breads and cereals, green vegetables, molasses
Selenium	• Immune mechanisms, mitochondrial adenosine triphosphate synthesis, cellular protection	• Seafood, meats, liver, eggs
Zinc	• Burn and wound healing, carbohydrate digestion, metabolism (carbohydrate, fat, protein), prostate gland function, reproductive organ growth and development, cell growth	• Liver, mushrooms, seafood, soybeans, spinach, meats

Digestion and absorption

Nutrients must be digested in the GI tract by enzymes that split large units into smaller ones. In this process, called *hydrolysis,* a compound unites with water and then splits into simpler compounds. The smaller units are then absorbed from the small intestine and transported to the liver through the portal venous system.

Carbohydrate digestion and absorption

Enzymes break down complex carbohydrates. In the oral cavity, *salivary amylase* initiates starch hydrolysis into disaccharides. In the small intestine, *pancreatic amylase* continues this process.

Splitting, hydrolyzing…

Disaccharide enzymes in the intestinal mucosa hydrolyze disaccharides into monosaccharides. Lactase splits the compound lactose into glucose and galactose, and sucrase hydrolyzes the compound sucrose into glucose and fructose.

… and movin' along

Monosaccharides, such as glucose, fructose, and galactose, are absorbed through the intestinal mucosa and are then transported through the portal venous system to the liver. There, enzymes convert fructose and galactose to glucose.

Ribonucleases and deoxyribonucleases break down nucleotides from deoxyribonucleic acid and ribonucleic acid into pentoses and nitrogen-containing compounds (nitrogen bases). Like glucose, these compounds are absorbed through the intestinal mucosa.

Gotta split! During hydrolysis, a compound unites with water and then splits into simpler compounds.

Protein digestion and absorption

Enzymes digest proteins by hydrolyzing the peptide bonds that link the amino acids of the protein chains. This process of hydrolyzation restores water molecules.

Gastric pepsin breaks proteins into:
- polypeptides
- pancreatic trypsin
- chymotrypsin
- carboxypeptidase, which converts polypeptides to peptides.

The breakdown lane

Intestinal mucosal peptidases break down peptides into their constituent amino acids. After being absorbed through the intestinal mucosa by active transport mechanisms, these amino acids travel through the portal venous system to the liver. The liver converts the amino acids not needed for protein synthesis into glucose.

Lipid digestion and absorption

Most fat digestion occurs in the small intestine. Pancreatic lipase breaks down fats and phospholipids into a mixture of glycerol, short- and long-chain fatty acids, and monoglycerides. The portal venous system then carries these substances to the liver. Lipase hydrolyzes the bonds between glycerol and fatty acids—a process that restores the water molecules released when the bonds were formed.

On a short leash...

Glycerol diffuses directly through the mucosa. Short-chain fatty acids diffuse into the intestinal epithelial cells and are carried to the liver via the portal venous system.

...or on a long one

Long-chain fatty acids and monoglycerides in the intestine dissolve in the bile salt micelles and then diffuse into the intestinal epithelial cells. There, lipase breaks down absorbed monoglycerides into glycerol and fatty acids. In the smooth endoplasmic reticulum of the epithelial cells, fatty acids and glycerol recombine to form fats.

Amino acids: Enter here and prepare to convert.

Chylomicrons

Along with a small amount of cholesterol and phospholipid, triglycerides are coated with a thin layer of protein to form lipoprotein particles called *chylomicrons*. Chylomicrons collect in the intestinal lacteals (lymphatic vessels) and are carried through lymphatic channels. After entering the circulation through the thoracic duct, they're distributed to body cells.

Stored away for later

In the cells, fats are extracted from the chylomicrons and broken down by enzymes into fatty acids and glycerol. Then they're absorbed and recombined in fat cells, reforming triglycerides for storage and later use.

Carbohydrate metabolism

Carbohydrates are the preferred energy fuel of human cells. Most of the carbohydrates in absorbed food is quickly catabolized for the release of energy.

Glucose to energy

All ingested carbohydrates are converted to glucose, the body's main energy source. Glucose not needed for immediate energy is stored as glycogen or converted to lipids.

Energy from glucose catabolism is generated in three phases:
1. glycolysis
2. Krebs cycle (also called the *citric acid cycle*)
3. the electron transport system. (See *Tracking the glucose pathway*, page 224.)

Glycolysis, which occurs in the cell cytoplasm, doesn't use oxygen. The other two phases, which occur in mitochondria, do use oxygen.

Glycolysis

Glycolysis refers to the process by which enzymes break down the 6-carbon glucose molecule into two 3-carbon molecules of pyruvic acid (pyruvate). Glycolysis yields energy in the form of adenosine triphosphate (ATP).

Cruising the glucose pathway

Next, pyruvic acid releases a carbon dioxide (CO_2) molecule and is converted in the mitochondria to a two-carbon acetyl fragment, which combines with a complex organic compound called *coenzyme A* (CoA) to form acetyl-CoA.

Krebs cycle

The second phase in glucose catabolism is the Krebs cycle. It's the pathway by which a molecule of acetyl-CoA is oxidized by enzymes to yield energy.

Carbons, carbons everywhere

The two-carbon acetyl fragments of acetyl-CoA enter the Krebs cycle by joining to the four-carbon compound oxaloacetic acid to form

This whole nutrition and metabolism thing is pretty simple; it's about getting energy to do things!

Now I get it!

Tracking the glucose pathway

Glucose catabolism generates energy in three phases: glycolysis, Krebs cycle, and the electron transport system. This flowchart summarizes the first two phases.

Glycolysis
Glycolysis, the first phase, breaks apart one molecule of glucose to form two molecules of pyruvate, which yields energy in the form of adenosine triphosphate and acetyl-coenzyme A (CoA).

Krebs cycle
The second phase, the Krebs cycle, continues carbohydrate metabolism. Fragments of acetyl-CoA join to oxaloacetic acid to form citric acid. The CoA molecule breaks off from the acetyl group and may form more acetyl-CoA molecules. Citric acid is first converted into intermediate compounds and then back into oxaloacetic acid. The Krebs cycle also liberates carbon dioxide.

Electron transport system
In the third phase of glucose catabolism, molecules on the inner mitochondrial membrane attract electrons from hydrogen atoms and carry them through oxidation-reduction reactions in the mitochondria. The hydrogen ions produced in the Krebs cycle then combine with oxygen to form water.

Glucose is the body's main energy source.

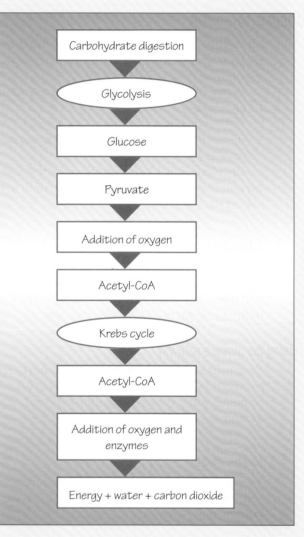

citric acid, a six-carbon compound. In this process, the CoA molecule detaches from the acetyl group, becoming available to form more acetyl-CoA molecules. Enzymes convert citric acid into intermediate compounds and eventually convert it back into oxaloacetic acid. Then the cycle can begin again.

In addition to liberating CO_2 and generating energy, each turn of the Krebs cycle releases hydrogen atoms, which are picked up by the coenzymes nicotinamide adenine dinucleotide (NAD) and flavin adenine dinucleotide (FAD).

The last step of carbohydrate catabolism is electron transport.

Electron transport system

The electron transport system is the last phase of carbohydrate catabolism. In this phase, carrier molecules on the inner mitochondrial membrane pick up electrons from the hydrogen atoms carried by NAD and FAD. (Each hydrogen atom contains a hydrogen ion and an electron.) These carrier molecules transport the electrons through a series of enzyme-catalyzed oxidation-reduction reactions in the mitochondria.

Oxygen attraction

Oxygen plays a crucial role by attracting electrons along the chain of carriers in the transport system. During oxidation, a chemical compound loses electrons; during reduction, it gains electrons. These reactions release the energy contained in the electrons and generate ATP.

After passing through the electron transport system, the hydrogen ions produced in the Krebs cycle combine with oxygen to form water.

Regulation of blood glucose levels

Because all ingested carbohydrates are converted to glucose, the body depends on the liver, muscle cells, and certain hormones to regulate blood glucose levels.

I convert glucose into glycogen or lipids.

Liver

When glucose levels exceed the body's immediate needs, hormones stimulate the liver to convert glucose into glycogen or lipids. Glycogen forms through glycogenesis; lipids form through lipogenesis.

Glucose shortage

When the blood glucose level drops excessively, the liver can form glucose by two processes:
- breakdown of glycogen to glucose through glycogenolysis
- synthesis of glucose from amino acids through gluconeogenesis.

Muscle cells

Muscle cells can convert glucose to glycogen for storage. However, they lack the enzymes to convert glycogen back to glucose when needed. During vigorous muscular activity, when oxygen requirements exceed the oxygen supply, muscle cells break down glycogen to yield lactic acid and energy. Lactic acid then builds up in the muscles, and muscle glycogen is depleted.

Glycogen returns

Some of the lactic acid diffuses from muscle cells, is transported to the liver, and is reconverted to glycogen. The liver converts the newly formed glycogen to glucose, which travels through the bloodstream to the muscles and reforms into glycogen.

Energize!

When muscle exertion stops, some of the accumulated lactic acid converts back to pyruvic acid. Pyruvic acid is oxidized completely to yield energy by means of the Krebs cycle and the electron transport system.

Hormones

Hormones regulate the blood glucose level by stimulating the metabolic processes that restore a normal level in response to blood glucose changes. (See *How insulin affects blood glucose level.*)

Sugar shift

Insulin, produced by the pancreatic islet cells, is the only hormone that significantly reduces the blood glucose level. In addition to promoting cell uptake and use of glucose as an energy source, insulin promotes glucose storage as glycogen (glycogenesis) and lipids (lipogenesis). Therefore, insulin production has widespread effects throughout the body.

Now I get it!

How insulin affects blood glucose level

Unlike most hormones, insulin tends to decrease the blood glucose level. It does this by helping glucose to enter cells, thereby promoting glycogenesis and stimulating glucose catabolism.

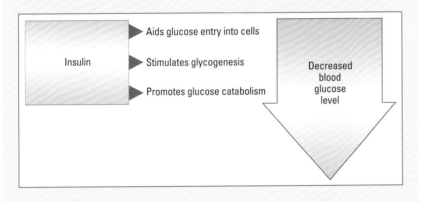

Protein metabolism

Proteins are absorbed as amino acids and carried by the portal venous system to the liver and then throughout the body by blood. Absorbed amino acids mix with other amino acids in the body's amino acid pool. These other amino acids may be synthesized by the body from other substances, such as *keto acids,* or they may be produced by protein breakdown.

Amino acid conversion

The body can't store amino acids; instead, it converts them to protein or glucose or catabolizes them to provide energy. Before these changes can occur, however, amino acids must be transformed by deamination or transamination.

Deamination

In *deamination,* an amino group ($-NH_2$) splits off from an amino acid molecule to form one molecule of ammonia and one of keto acid. Most of the ammonia is converted to urea and excreted in urine.

Transamination

In *transamination*, an amino group is exchanged for a keto group in a keto acid through the action of transaminase enzymes. During this process, the amino acid is converted to a keto acid and the original keto acid is converted to an amino acid.

Amino acid synthesis

Proteins are synthesized from 20 amino acids from the body's amino acid pool. (See *Essential and nonessential amino acids*.)

Power station

Amino acids not used for protein synthesis can be converted to keto acids and metabolized by the Krebs cycle and the electron transport system to produce energy.

Essential and nonessential amino acids

Amino acids are the structural units of proteins. They're classified as essential or nonessential based on whether the human body can synthesize them. The 9 essential amino acids that can't be synthesized must be obtained from the diet. The other 11 can be synthesized and are therefore nonessential in the diet; however, they're needed for protein synthesis.

Essential
- Histidine
- Isoleucine
- Leucine
- Lysine
- Methionine
- Phenylalanine
- Threonine
- Tryptophan
- Valine

Nonessential
- Alanine
- Arginine
- Asparagine
- Aspartic acid
- Cystine
- Glutamine
- Glycine
- Hydroxyproline
- Proline
- Serine
- Tyrosine

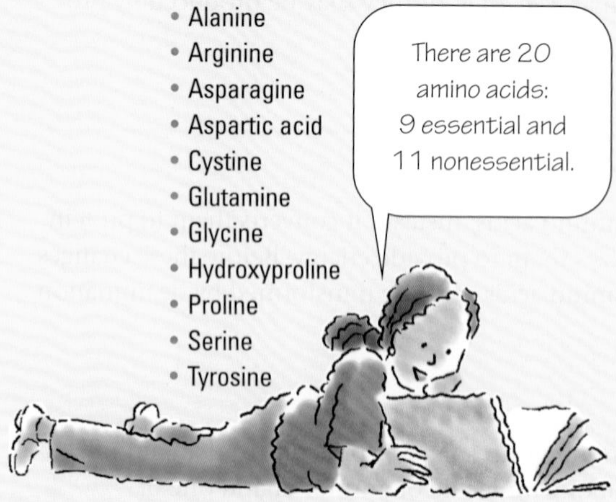

There are 20 amino acids: 9 essential and 11 nonessential.

Fat chance

Amino acids not used for protein synthesis may be converted to pyruvic acid and then to acetyl-CoA. The acetyl-CoA fragments condense to form long-chain fatty acids—a process that's the reverse of fatty acid breakdown. These fatty acids then combine with glycerol to form fats.

Going glucose

Amino acids can also be converted to glucose. They're first converted to pyruvic acid, which may then be converted to glucose.

Lipid metabolism

Lipids are stored in adipose tissue within cells until they're required for use as fuel. When needed for energy, each fat molecule is hydrolyzed to glycerol and three molecules of fatty acids. Glycerol can be converted to pyruvic acid and then to acetyl-CoA, which enters the Krebs cycle.

Ketone body formation

The liver normally forms ketone bodies from acetyl-CoA fragments, derived largely from fatty acid catabolism. Acetyl-CoA molecules yield three types of ketone bodies: acetoacetic acid, beta-hydroxybutyric acid, and acetone.

Acetoacetic acid results from the combination of two acetyl-CoA molecules and the subsequent release of CoA from these molecules.

Beta-hydroxybutyric acid forms when hydrogen is added to the oxygen atom in the acetoacetic acid molecule. The term *beta* indicates the location of the carbon atom containing the OH group.

Acetone forms when the COOH group of acetoacetic acid releases CO_2. Muscle tissue, brain tissue, and other tissues oxidize these ketone bodies for energy.

Along with other tissues, I oxidize ketone bodies for energy.

Excessive ketone formation

Under certain conditions, the body produces more ketone bodies than it can oxidize for energy. Such conditions include fasting, starvation, and uncontrolled diabetes (in which the body can't break down glucose). The body must then use fat, rather than glucose, as its primary energy source.

Ketone cops

Use of fat instead of glucose for energy leads to an excess of ketone bodies. This condition disturbs the body's normal acid-base balance and homeostatic mechanisms, leading to ketosis.

Lipid formation

Excess amino acids can be converted to fat through keto acid–acetyl-CoA conversion. Glucose may be converted to pyruvic acid and then to acetyl-CoA, which is converted into fatty acids and then fat (in much the same way that amino acids are converted into fat). (See *Nutrition-related changes with aging.*)

Senior moment

Nutrition-related changes with aging

As a person ages, caloric needs decrease. Protein, vitamin, and mineral requirements usually remain the same throughout life. The body's ability to process these nutrients, however, is also affected by the aging process.

Physiologic changes
Diminished intestinal motility typically accompanies aging and may cause constipation. Physical inactivity, emotional stress, medications, and nutritionally inadequate diets of soft, refined foods that are low in dietary fiber can also cause constipation. Laxative abuse results in the rapid transport of food through the GI tract, decreasing digestion and absorption. Fecal incontinence may also occur in elderly patients.

Other physiologic changes that can affect nutrition in an older patient include:
• decreased renal function, causing greater susceptibility to dehydration and formation of renal calculi
• loss of calcium and nitrogen (in people who aren't ambulatory)
• diminished enzyme activity and gastric secretions

• reduced pepsin and hydrochloric acid secretion, which tends to diminish the absorption of calcium and vitamins B_1 and B_2
• decreased salivary flow and diminished sense of taste, which may reduce the appetite and increase a person's consumption of sweet and spicy foods
• diminished intestinal motility and peristalsis of the large intestine
• thinning of tooth enamel, causing teeth to become more brittle
• decreased biting force
• diminished gag reflex.

Affecting factors
Nutritional status can be affected by such socioeconomic and psychological factors as loneliness, decline of the older person's importance in the family, susceptibility to nutritional quackery, and lack of money or transportation, which limit access to nutritious foods. In addition, some conditions that are common among older people can affect mobility and, therefore, the ability to obtain or prepare food or feed oneself.

Quick quiz

1. Which type of nutrient yields 9 kcal/g when used for energy?
   A. Proteins
   B. Carbohydrates
   C. Lipids
   D. Vitamins

Answer: C. Lipids are a concentrated form of fuel and yield 9 kcal/g.

2. Which hormone decreases the blood glucose level?
   A. Epinephrine
   B. Cortisol
   C. Insulin
   D. Glycogen

Answer: C. Insulin is the only hormone that significantly reduces blood glucose. It does so by aiding glucose entry into cells, thereby stimulating glycogenesis and promoting glucose catabolism.

3. Essential amino acids are:
   A. organic compounds that are needed in small amounts for normal metabolism, growth, and development.
   B. organic compounds that don't dissolve in water but do dissolve in alcohol and other organic solvents.
   C. the structural unit of protein that doesn't need to be obtained from the diet.
   D. the structural unit of protein that must be obtained from the diet.

Answer: D. Essential amino acids can't be synthesized in the body and, therefore, must be obtained from the diet.

4. Which vitamin is involved in prothrombin synthesis and other blood-clotting factors?
   A. Vitamin K
   B. Vitamin E
   C. Vitamin B_{12}
   D. Vitamin B_6

Answer: A. Vitamin K is involved in liver synthesis of prothrombin and other blood-clotting factors.

Scoring

⭐⭐⭐ If you answered all four questions correctly, hooray! You've digested a healthy dose of dense clinical material.

⭐⭐ If you answered three questions correctly, remarkable! You'll soon be a master of metabolism.

⭐ If you answered fewer than three questions correctly, get energized! Just a few more quick quizzes to go!

Just for fun!

Unscramble the words on the left to discover the names of three major types of nutrients. Then draw lines to link each nutrient to the correct information. Note that some information may apply to more than one nutrient.

| OPENSTIR |
| _ _ _ _ _ _ _ |

| SLIDPI |
| _ _ _ _ _ _ |

| ACABREDHOSTRY |
| _ _ _ _ _ _ _ _ _ _ _ _ _ |

A. Concentrated form of fuel
B. Used mainly for growth and repair of body tissues
C. Yield 9 kcal/g when used for energy
D. Organic compounds composed of carbon, hydrogen, and oxygen
E. Complex nitrogenous organic compounds containing amino acid chains
F. Yield 4 kcal/g when used for energy
G. Organic compounds that don't dissolve in water but do dissolve in alcohol and other organic solvents
H. The body's primary energy source
I. Composed of amino acids
J. Classified as monosaccharides, disaccharides, and polysaccharides
K. Major types include fats, phospholipids, and steroids

Answer: PROTEINS: B, E, F, I; LIPIDS: A, C, G, K; CARBOHYDRATES: D, F, H, J

Selected References

Academy of Nutrition and Dietetics. (2016). *Vitamins and minerals.* Retrieved from http://www.eatright.org, on June, 2016.

Smolin, L. A., & Grosvenor, M. B. (2013). *Nutrition: Science and applications* (3rd ed.). Wiley Hoboken, NJ.

United States Food and Drug Administration. (2016). *Fortify your knowledge about vitamins.* Retrieved from http://www.fda.gov/ForConsumers/ConsumerUpdates/ucm118079.htm, on June, 2016.

United States Department of Agriculture. (2016). Interactive tools. Retrieved from https://fnic.nal.usda.gov/dietary-guidance/interactive-tools, on June, 2016.

Chapter 14

Urinary system

Just the facts

In this chapter, you'll learn:

◆ major structures of the urinary system

◆ functions of the kidneys, ureters, bladder, and urethra

◆ the way in which urine is formed

◆ role of hormones in the urinary system.

Structures of the urinary system

The urinary system consists of:
- two kidneys
- two ureters
- the urinary bladder
- the urethra.

Working together, these structures remove and excrete wastes from the body via urine; help to govern fluid, electrolyte, and acid-base balance; have a role in the production of red blood cells; and assist in blood pressure control.

> Think of us as the body's plumbers. Together with the ureters, bladder, and urethra, we do everything from removing wastes…

> …to balancing fluids and electrolytes, to keeping blood pressure in check.

Kidneys

The *kidneys* are bean-shaped organs approximately 11 cm long by 6 cm wide by 3 cm thick. They lie in the posterior abdominal wall on either side of the vertebral column. Their major functions are excretion, elimination, and homeostasis of blood plasma. Each kidney consists of three regions: the renal cortex, renal medulla, and renal pelvis.

Renal cortex

The *renal cortex* is the outermost region of the kidney and lies between the renal capsule and renal medulla. It contains blood-filtering mechanisms (renal corpuscles and tubules), collection ducts, and a large vascular network. It is relatively smooth except for the renal columns (cortical columns) that project from the renal cortex into the renal medulla to separate the renal pyramids.

Renal medulla

The *renal medulla* is the inner region of the kidney. It is composed of conical-shaped masses of tissue called the renal pyramids (medullary); each kidney contains 6 to 18 renal pyramids—wedges composed of parallel bundles of tubules and capillaries. A major tubule, the loop of Henle, extends from the renal cortex into the renal medulla. Each pyramid is separated by columns from the cortex. The broad portion of each pyramid lies toward the cortex and the tapered portion of each pyramid (the papilla) toward the renal pelvis.

Renal pelvis (See *A close look at the urinary system*)

The *renal pelvis* is a funnel-shaped tube with branching extensions called calyces (singular calyx). The minor calyces form larger 2 to 3 major calyces. These calyces are cuplike containers that collect urine from the papillae of each renal pyramid and drain into the renal pelvis, which then transports urine to the ureter. The calyces, renal pelvis, and ureter are all lined with smooth muscle, and the urine is moved via peristaltic movement (Faiz, Blackburn, & Moffat, 2011; Marieb & Hoehn, 2014; Netter, Hansen, & Lambert, 2005; Peate & Nair, 2015; Shier, Butler, & Lewis, 2015; Waugh & Grant, 2014).

Safe placement

The kidneys are retroperitoneal organs and, as such, are protected in front by the contents of the abdomen. In addition, they receive some protection from the lower portion of the rib cage and behind from the muscles attached to the vertebral column. The major protective mechanism is the layer of fat each kidney lies embedded in. The kidney and fatty tissue are then surrounded by protective fibrous capsule of connective tissue called the *renal fascia*. The right kidney lies slightly lower than the left.

Adrenal influence

On top of each kidney lies an adrenal gland. These glands are affected by the release of renin from the kidneys and, in turn, affect the renal system by secreting aldosterone, which directly influences the amount of sodium and water reabsorbed by the kidneys resulting in changes in blood pressure.

Blood and nerve supply

The kidneys receive approximately one-quarter of the total cardiac output each minute (1,200 mL). This volume is supplied by the renal arteries. The right and left renal arteries branch off of the abdominal aorta and then subdivide into five segmental arteries and finally branch into several interlobar arteries prior to entering the kidney. Approximately 90% of the blood that enters the kidney perfuses the renal cortex. Blood exits the kidneys via the renal veins and empties into the inferior vena cava.

The nerve supply to each kidney and its corresponding ureter is supplied via the renal plexus, a network of autonomic fibers and ganglia. The renal plexus is supplied by the sympathetic fibers from the inferior thoracic and first lumbar splanchnic nerves (Faiz et al., 2011; Marieb & Hoehn, 2014; Shier et al., 2015; Waugh & Grant, 2014).

All in a day's work

Together, these tissues allow the kidneys to perform their many functions, including:

- excretion and elimination of wastes (in the form of urine)
- blood filtration (by regulating chemical composition and blood volume)
- homeostasis (fluid-electrolyte and acid-base balances)
- production and release of renin to promote angiotensin II activation and aldosterone production in the adrenal gland
- production of erythropoietin (a hormone that stimulates red blood cell [RBC] production) and enzymes (such as renin, which governs blood pressure and kidney function)
- conversion of vitamin D to a more active form.

The nephron

The *nephron* serves as the basic structural and functional unit of the kidney. (See *Structure of the nephron.*) Each kidney contains over 1 million of these processing units and thousands of collecting ducts.

Zoom in

A close look at the urinary system

The kidneys are located in the superior lumbar region, with the right kidney situated slightly lower than the left to make room for the liver, which is just above it. The position of the kidneys shifts somewhat with changes in body position. Surrounding the kidneys are the fibrous capsule, perirenal fat capsule, and renal fasciae.

Blood's cleansing journey

The kidneys receive waste-filled blood from the renal artery, which branches off the abdominal aorta. After passing through a complicated network of smaller blood vessels and nephrons, the filtered blood returns to the circulation by way of the renal vein, which empties into the inferior vena cava.

Continuing the cleanup

The kidneys excrete waste products that the nephrons remove from the blood; these excretions combine with other waste fluids (such as urea, creatinine, phosphates, and sulfates) to form urine. An action called *peristalsis* (the circular contraction and relaxation of a tube-shaped structure) passes the urine through the ureters and into the urinary bladder. When the bladder has filled, nerves in the bladder wall relax the sphincter. In conjunction with a voluntary stimulus, this relaxation causes urine to pass into the urethra for elimination from the body.

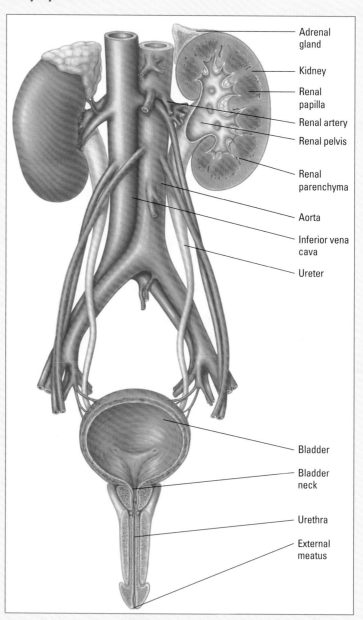

Adrenal gland

Kidney

Renal papilla

Renal artery

Renal pelvis

Renal parenchyma

Aorta

Inferior vena cava

Ureter

Bladder

Bladder neck

Urethra

External meatus

Nephrons perform two main functions:

- mechanically filtering fluids, wastes, electrolytes, acids, and bases into the tubular system
- selectively reabsorbing and secreting ions, allowing precise control of fluid and electrolyte balance.

Each nephron consists of a renal corpuscle and a renal tubule.

Renal corpuscle

All of the renal corpuscles are located in the renal cortex and consist of a cluster of fenestrated (porous) capillaries called a *glomerulus*, situated inside a hollow collecting duct called the *glomerular capsule* or *Bowman's capsule*. The glomerular capsule is also continuous with its renal tubule. This is where the blood flowing into the kidney from the arteries is turned into a cell- and protein-free filtrate. The renal tubules are where the body selectively reclaims the substances that need to be reabsorbed back into the blood. Approximately 99% of these substances are reabsorbed. Anything that is not reabsorbed becomes urine.

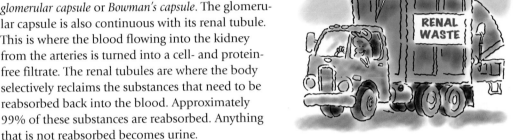

> The nephrons mechanically process fluids, wastes, electrolytes, acids, and bases into the tubular system.

RENAL WASTE

Renal tubule

Renal tubules begin in the renal cortex, pass into the medulla, and return to the cortex. Each tubule is approximately 3 cm long and has three parts:

- Proximal convoluted tubule—the coiled section that exits the glomerular capsule and has freely permeable cell membranes. This allows reabsorption of nearly all the filtrate's glucose, amino acids, metabolites, and electrolytes into nearby capillaries as well as allowing for the circulation of large amounts of water.
- Nephron loop (loop of Henle)—hairpin loop with descending and ascending regions, together with their accompanying blood vessels and collecting tubules, form the renal pyramids in the medulla.
- Distal convoluted tubule—winds until it empties into a collecting duct. Each portion has a different processing role.

The glomeruli and proximal and distal tubules of the nephron are located in the renal cortex. Filtrates pass through the length of the tubules, from the proximal to the distal portion, moving from the glomerulus to the collecting duct.

Collecting duct

The filtrate from the nephrons is channeled to collecting ducts in the medullary pyramids.

There are two types of cell in each collecting duct: *principal* and *intercalated cells*. The principal cells maintain the water and sodium, and the intercalated cells maintain acid-base balance. The collecting ducts fuse together in the renal pelvis and open into the minor calyces to deliver urine.

Zoom in

Structure of the nephron

The *nephron* is the kidney's basic functional unit and the site of urine formation. Each nephron is composed of a renal corpuscle and renal tubule. The renal artery, a large branch of the abdominal aorta, carries blood to each kidney. Blood flows through the interlobular artery to the afferent arteriole, which conveys blood to the glomerulus. Blood is forced through the membranes of the glomerulus into the efferent arteriole and into the peritubular capillaries, venules, and the interlobular vein. The peritubular capillary network of vessels then supplies blood to the tubules of the nephron.

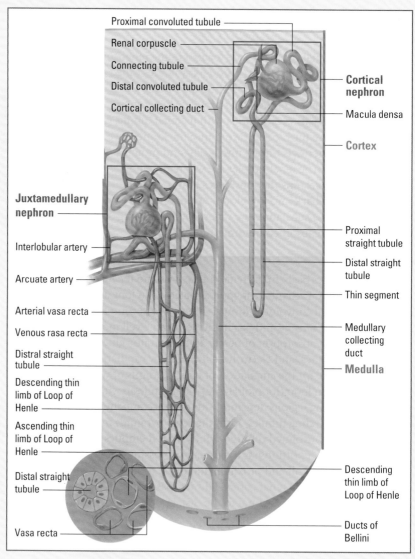

Proximal convoluted tubule

Renal corpuscle

Connecting tubule

Distal convoluted tubule

Cortical collecting duct

Cortical nephron

Macula densa

Cortex

Juxtamedullary nephron

Interlobular artery

Arcuate artery

Arterial vasa recta

Venous rasa recta

Distral straight tubule

Descending thin limb of Loop of Henle

Ascending thin limb of Loop of Henle

Distal straight tubule

Vasa recta

Proximal straight tubule

Distal straight tubule

Thin segment

Medullary collecting duct

Medulla

Descending thin limb of Loop of Henle

Ducts of Bellini

Senior moment

Urinary changes with aging

As a person ages, changes in the kidneys and bladder can affect urinary system function.

Kidneys

After age 40, kidney function may diminish; if the person lives to age 90, it may decrease by as much as 50%. Age-related changes in kidney vasculature that disturb glomerular hemodynamics result in a decline in glomerular filtration rate. Reduced cardiac output and age-related atherosclerotic changes cause kidney blood flow to decrease by 53%. In addition, tubular reabsorption and renal concentrating ability decline because the size and number of functioning nephrons decrease. Also, as blood levels of aldosterone and renin fall, the kidneys are less responsive to antidiuretic hormone.

Bladder

As a person ages, bladder muscles weaken. This may lead to incomplete bladder emptying and chronic urine retention—predisposing the bladder to infection.

And the rest

Other age-related changes that affect renal function include diminished kidney size, impaired renal clearance of drugs, reduced bladder size and capacity, and decreased renal ability to respond to variations in sodium intake. By age 70, blood urea nitrogen levels rise by 21%. Residual urine, frequency of urination, and nocturia also increase with age.

Time to concentrate

There is a specialized group of nephrons, called *juxtamedullary* nephrons, near the junction of the cortex and medulla. Their long loops have an important role in the kidney's ability to produce concentrated urine. By the time the filtrate enters the descending limb of the loop of Henle, approximately 70% of its water content has been reabsorbed. At this point, the filtrate contains a high concentration of salts, chiefly sodium. As the filtrate moves deeper into the medulla and the loop of Henle, osmosis draws even more water into the extracellular spaces, further concentrating the filtrate.

When the filtrate enters the distal tubule, its concentration is readjusted by the secretion of ions (K^+, H^+, HCO_3), reabsorption of sodium, or conservation of water, depending on the composition of the interstitial fluid. The concentration can be regulated until the filtrate leaves the collection duct as urine. (*See Urinary changes with aging.*)

Ureters

The *ureters* are fibromuscular tubes that convey urine from the kidneys to the bladder. Each ureter descends from the kidneys and enters obliquely through the posterior bladder wall, preventing backflow of urine. Because the left kidney is higher than the right kidney, the left ureter is usually slightly longer than the right ureter. (See *Three layers of the ureter*.)

Urine does not travel from the kidneys to the bladder by gravity alone. As urine enters the ureter, it stimulates the muscularis to contract. Peristaltic waves occurring one to five times each minute channel urine along the ureters toward the bladder.

Urinary bladder

The *urinary bladder* is a hollow, smooth, collapsible, muscular organ in the pelvis. It lies retroperitoneally in the pelvic cavity and posterior to the *symphysis pubis* (the joint between the two pubic bones). Its function is to store urine. In a normal adult, a full bladder holds approximately 500 mL. If the bladder is nearing capacity, it rises superiorly in the pelvis and can be palpated above the *symphysis pubis*. The maximum capacity of the bladder is 800 to 1,000 mL, and if distended past this point, it can burst.

The interior of the bladder contains openings for two ureters and a urethra. The smooth, triangular region of the bladder created by these openings is called the *trigone*. Clinically, this is where infection is likely to persist. The bladder wall consists of three layers:
- the *mucosa*—made of transitional epithelium, the innermost layer
- the muscular layer (*detrusor*)—combination of circular and longitudinal smooth muscle
- the *adventitia*—fibrous external layer covering everything except the superior surface (which is covered by the peritoneum).

Urethra

The *urethra* is a thin, muscular tube that channels urine from the bladder to the outside of the body. Urination results from involuntary (reflex) and voluntary (learned or intentional) processes. At the junction of the urethra and the bladder, the detrusor muscle thickens to form the *internal urethral sphincter*. The internal sphincter is controlled by the autonomic nervous system and is involuntary. It prevents leakage of urine when not actively urinating. Once the urethra passes through the urogenital diaphragm, it is surrounded by a second sphincter, called the *external urethral sphincter*. It is made up of skeletal muscle, and in conjunction with the levator ani muscle located in the pelvic floor, it serves as a voluntary sphincter. When urine fills the

Three layers of the ureter

Each ureter is composed of three layers of tissue:
- The *mucosa*, the innermost layer, contains the transitional epithelium.
- The *muscularis* contains longitudinal and circular smooth muscle layers.
- The *adventitia*, an external surface of fibrous connective tissue.

Urination is the result of two processes—one voluntary and one involuntary.

bladder, parasympathetic nerve fibers in the bladder wall cause the bladder to contract and the *internal sphincter* (located at the internal urethral orifice) to relax. The cerebrum, in a voluntary reaction, then causes the external sphincter to relax and urination to begin. This is called the *micturition* reflex.

Female variations

In the female, the urethra is 3 to 4 cm long and is bound to the anterior wall of the vagina by fibrous connective tissue. The urethra connects the bladder with an external opening, or *urethral meatus,* located anterior to the vaginal opening and posterior to the clitoris.

Male variations

In the male, the urethra is approximately 20 cm in length and is made up of three regions: the 2.5-cm *prostatic urethra,* the 2-cm *membranous urethra,* and the 15-cm *spongy urethra.* The prostatic urethra passes vertically through the *prostate* gland. The membranous urethra extends through the urogenital diaphragm. The spongy urethra extends from the diaphragm through the penis to the external urethral orifice. The male urethra serves as a passageway for semen as well as urine.

Urine formation

Urine formation is one of the main functions of the urinary system. Urine formation results from three processes that occur in the nephrons: glomerular filtration, tubular reabsorption, and tubular secretion. (See *How the kidneys form urine.*)

Urine is typically clear, yellow, and slightly acidic. It is 95% water, with a specific gravity of 1.001 to 1.035. Solutes in urine are typically urea, uric acid, and creatinine (nitrogenous waste) and various ions: sodium, potassium, calcium, magnesium, sulfates, phosphates, bicarbonates, uric acid, ammonium ions, and *urobilinogen* (a derivative of bilirubin resulting from the action of intestinal bacteria). A few leukocytes and RBCs (and, in males, some spermatozoa) may enter the urine as it passes from the kidney to the ureteral orifice. When a person is taking drugs that are normally excreted in urine, the urine contains those substances as well.

Kidneys in charge

The kidneys can vary the amount of substances reabsorbed and secreted in the nephrons, changing the composition of excreted urine.

Controlling the flow

Total daily urine output averages 800 to 2,000 mL, with a normal fluid intake of 2 L/day. The output varies with fluid intake, exercise, age, and climate. For example, after drinking a large volume of fluid, a person's urine output increases as the body rapidly excretes excess water. If a person restricts or decreases water intake or ingests excessive amounts of sodium, urine output decreases as the body retains water to restore normal fluid concentration.

We'd better get busy! We process up to 300 glasses of water every day!

Hormones and the urinary system

Hormones play a major role in the urinary system, including helping the body to manage tubular reabsorption and secretion. Hormones affecting the urinary system include:

- renin
- antidiuretic hormone
- angiotensin I
- angiotensin II
- aldosterone
- erythropoietin.

Now I get it!

How the kidneys form urine

Urine formation occurs in three steps: glomerular filtration, tubular reabsorption, and tubular secretion.

Step 1: Filtrate
Blood flowing through the arterial network into the kidney is passively forced through the 3-part filtration membrane of the glomerulus by hydrostatic pressure. This process blocks large molecules like blood cells and plasma protein from entering the glomerular capsule and helps to maintain osmotic pressure in the surrounding capillaries. When blood and protein spill into the urine, it usually indicates a problem with glomerular filtration.

Step 2: Reabsorb
Approximately 99% of the filtrate is selectively reabsorbed back into the blood as it moves through the renal tubules and collection ducts. Depending on the substance, transport may be active (requiring energy) or passive. Active tubular reabsorption is the active transport of sodium (Na^+), potassium (K^+), and glucose across the membranes

of the tubule back into the peritubular capillaries. Passive reabsorption of water (H_2O) in the tubule is facilitated by osmosis based on the osmotic gradient caused by the movement of sodium and other solutes. The presence of antidiuretic hormone causes H_2O reabsorption in the collecting duct, while aldosterone regulates the absorption of sodium in the collecting duct. The body's total plasma volume circulates through the renal tubules approximately three times per hour. So, if the renal tubules did not reabsorb the majority of the fluids filtering through the kidney, it would only take about 30 minutes before the entirety of the plasma volume became urine.

Step 3: Secrete
Tubular secretion is the reverse of tubular reabsorption; selected substances are moved from the peritubular capillaries through the cells of the tubule into the filtrate. The proximal convoluted tubule is the main site of secretion for all substances with the exception of potassium.

How the kidneys form urine *(continued)*

Secretion of potassium is done via aldosterone regulation in the distal convoluted tube and collecting duct. The process of secretion allows for elimination of certain drugs and metabolites, unwanted substances that have been passively reabsorbed, excreting excess potassium, and maintaining blood pH.

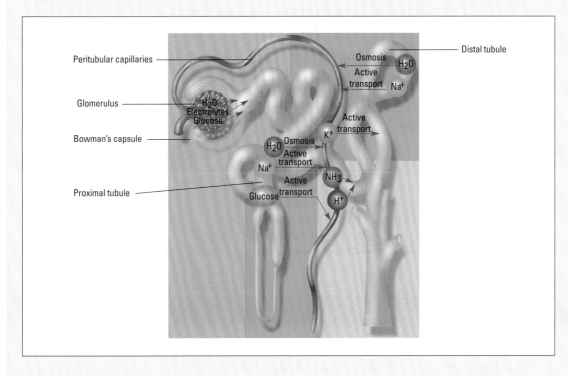

Antidiuretic hormone

Antidiuretic hormone (ADH) is produced by the hypothalamus and stored in the posterior pituitary. It regulates fluid balance and urine output. High levels of ADH increase water absorption and urine concentration, whereas lower levels of ADH decrease water absorption and dilute urine. (See *How antidiuretic hormone works.*)

Renin-angiotensin-aldosterone system

Renin is an enzyme that's secreted by juxtaglomerular cells in the kidneys and circulated in the blood. It then converts angiotensinogen released by the liver to into *angiotensin I*. As it circulates through the lungs, angiotensin I is converted into *angiotensin II* by *angiotensin-converting enzyme*. Angiotensin II exerts a powerful constricting effect on the arterioles. In this way, it can raise blood pressure. Angiotensin II also stimulates secretion of aldosterone from the adrenal glands. Aldosterone influences fluid balance by regulating absorption of sodium and water in the renal tubules and collecting ducts.

BP assist

Aldosterone, a mineralocorticoid, facilitates tubular reabsorption of fluid by regulating sodium retention and potassium secretion via epithelial cells in the renal tubules.

If serum potassium levels rise, there is inadequate blood flow to the kidneys, or blood pressure drops, the adrenal cortex responds by increasing aldosterone secretion. This, in turn, causes sodium retention, thereby raising blood pressure. (See *The renin-angiotensin-aldosterone system.*)

The best defense

The primary function of the renin-angiotensin system is to serve as a defense mechanism, maintaining blood pressure in situations such as hemorrhage and extreme sodium depletion. Low blood pressure and low levels of sodium passing through the kidneys are two of the three factors that stimulate the kidneys to release renin. (The third is stimulation of the sympathetic nervous system.)

High levels of ADH increase water reabsorption and urine concentration, whereas lower levels of ADH decrease water reabsorption and dilute urine.

Now I get it!

How antidiuretic hormone works

Antidiuretic hormone (ADH) regulates fluid balance in four steps.

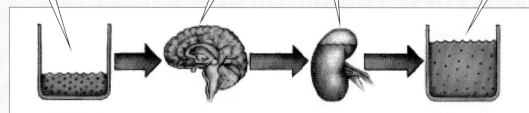

1. Low blood volume and increased serum osmolality are detected by the hypothalamus, which stimulates the pituitary gland.

2. The pituitary gland secretes ADH into the bloodstream.

3. ADH travels through the blood to the kidneys and causes the tubules and collecting ducts to become more permeable to water.

4. Water reabsorption increases, increasing plasma volume and decreasing serum osmolality.

Erythropoietin

The kidneys secrete the hormone *erythropoietin* in response to low arterial oxygen levels. This hormone travels to the bone marrow, where it stimulates increased production of red blood cells to improve the oxygen-carrying capacity of the blood.

A balancing act

The kidneys also regulate calcium and phosphorus balance by filtering and reabsorbing approximately half of unbound serum calcium. In addition, the kidneys activate vitamin D_3, a compound that promotes intestinal calcium absorption and regulates phosphate excretion.

Now I get it!

The renin-angiotensin-aldosterone system

The renin-angiotensin-aldosterone system regulates the body's sodium and water levels and blood pressure. Juxtaglomerular cells (1) near the glomeruli in each kidney secrete the enzyme renin into the blood.

Renin circulates throughout the body and converts angiotensinogen, made in the liver (2), into angiotensin I. In the lungs (3), angiotensin I is converted by hydrolysis to angiotensin II. Angiotensin II acts on the adrenal cortex (4) to stimulate production of the hormone aldosterone. Aldosterone acts on the juxtaglomerular cells to increase sodium and water retention and to stimulate or depress further renin secretion, completing the feedback system that automatically readjusts homeostasis.

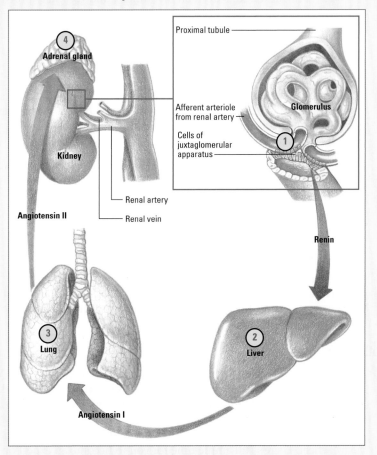

Quick quiz

1. In a normal adult, a full bladder is approximately:
    A. 50 to 100 mL.
    B. 200 to 300 mL.
    C. 500 to 600 mL.
    D. 700 to 900 mL.

Answer: C. In a normal adult, a full bladder ranges from 500 to 600 mL.

2. The left ureter is slightly longer than the right because the:
    A. left kidney is higher than the right.
    B. right kidney is higher than the left.
    C. left kidney performs more functions.
    D. left ureter has a three-layered wall.

Answer: A. The left kidney is slightly higher than the right kidney. Therefore, the left ureter needs to be longer to reach the bladder.

3. Urination results from an involuntary and voluntary process. This process is called the:
    A. kidney process.
    B. glomerular filtration rate.
    C. prostate reflex.
    D. micturition reflex.

Answer: D. The micturition reflex is the signal system that occurs when urine fills the bladder. Parasympathetic nerve fibers in the bladder wall cause the bladder to contract and the internal sphincter to relax, which is followed by a voluntary relaxation of the external sphincter.

4. A person on a new health regimen has begun to drink at least 8 oz (236 mL) of water six to eight times per day. The kidneys react to this change by:
    A. producing aldosterone.
    B. secreting renin.
    C. increasing urine output.
    D. secreting erythropoietin.

Answer: C. Urine output typically varies with fluid intake and climate. After ingestion of a large volume of fluid, urine output increases as the body rapidly excretes excess water.

Scoring

✰✰✰   If you answered all four questions correctly, congratulations! You've got the kidneys (and ureters, bladder, and urethra) down cold.

✰✰   If you answered three questions correctly, not bad! You're learning to navigate the urinary system.

✰   If you answered fewer than three questions correctly, be of good cheer. There's still ample time for you to become an expert on anatomy and physiology!

Just for fun!

Unscramble the following words to discover the names of the layers of the GI tract wall. Then draw lines to connect each layer with its particular characteristics.

LEARN TREXOC
_ _ _ _ _
_ _ _ _ _ _

NEARL LAUDELM
_ _ _ _ _
_ _ _ _ _ _ _

REALN LEVIPS
_ _ _ _ _
_ _ _ _ _ _

A. Inner region of the kidney
B. Contains blood-filtering mechanisms
C. Functions as the kidney's collecting chambers
D. Outer region of the kidney
E. Protected by a fibrous capsule and layers of fat
F. Middle region of the kidney
G. Receives urine through the major calyces

Answer: RENAL CORTEX: B, D, E; RENAL MEDULLA: C, F; RENAL PELVIS: A, G

Selected References

Considine, R. V., Rhoades, R. A., & Bell, D. R. (2009). Chapter 38: Fertilization, pregnancy and fetal differentiation. In *Medical physiology: Principles for clinical medicine* (3rd ed.). Baltimore, MD: Lippincott Williams & Wilkins.

Cooper, T. G.; World Health Organization (WHO). (2010). WHO reference values for human semen characteristics. *Human Reproduction Update, 16*(5), 559.

Faiz, O., Blackburn, S., & Moffat, D. (2011). *Anatomy at a glance* (3rd ed.). Oxford, UK: Wiley-Blackwell.

Marieb, E. N., & Hoehn, K. N. (2014). *Human anatomy and physiology: Pearson New International Edition* (9th ed.). Essex, UK: Pearson Education Limited.

Netter, F. H., Hansen, J. T., & Lambert, D. R. (2005). *Netter's clinical anatomy.* Carlstadt, NJ: Icon Learning Systems.

Oxford Medical Publications; Wilkins, R., Cross, S., Megson, I., & Meredith, D. (Eds.). (2011). *Oxford handbook of medical sciences.* Oxford, UK: Oxford University Press.

Peate, I., & Nair, M. (2015). *Anatomy and physiology for nurses at a glance.* Oxford, UK: Wiley-Blackwell.

Porter, R. S., Kaplan, J. L., & Merck & Co. (2011). *The Merck manual of diagnosis and therapy.* Whitehouse Station, NJ: Merck Sharp & Dohme Corp.

Shier, D., Butler, J., & Lewis, R. (2015). *Hole's essentials of human anatomy & physiology* (12th ed.). New York, NY: McGraw-Hill Education.

Smith, L. B., & Walker, W. H. (2014). The regulation of spermatogenesis by androgens. *Seminars in Cell & Developmental Biology, 30*, 2–13. doi: 10.1016/j.
  semcdb.2014.02.012.

Standring, S., & Gray, H. (2008). *Gray's anatomy: The anatomical basis of clinical practice.*
  Edinburgh, Scotland: Churchill Livingstone/Elsevier.

Waugh, A., & Grant, A. (2014). *Ross and Wilson: Anatomy and physiology in health and
  illness* (12th ed.). Edinburgh, Scotland: Churchill Livingstone/Elsevier.

Chapter 15

Fluids, electrolytes, acids, and bases

Just the facts

In this chapter, you'll learn:

♦ the way in which fluids are distributed throughout the body
♦ the kidneys' role in electrolyte balance
♦ the body's way of compensating for acid-base imbalances
♦ the major acid-base imbalances.

Fluid balance

The health and *homeostasis* (equilibrium of the various body functions) of the human body depend on *fluid, electrolyte,* and *acid-base balance*. Factors that disrupt this balance, such as surgery, illness, and injury, can lead to potentially fatal changes in metabolic activity. (See *How the body gains and loses fluids,* page 250.)

The four fluids

Body fluid is made up of water containing *solutes,* or dissolved substances, that are necessary for physiologic functioning. Solutes include electrolytes, glucose, amino acids, and other nutrients. There are four types of body fluids:

• *Intracellular fluid (ICF)* is found within the individual cells of the body.
• *Intravascular fluid (IVF)*, also known as plasma, is found within the blood vessels and the lymphatic system.

Fluids, electrolytes, acids, and bases— it's a balancing act to keep the body functioning correctly.

Body shop

How the body gains and loses fluids

Each day, the body takes in fluid from the GI tract (in foods, liquids, and water of oxidation) and loses fluids through the skin, lungs, intestines (stool), and urinary tract (urine). This illustration shows the primary sites involved in fluid gains and losses as well as the amount of normal daily fluid intake and output.

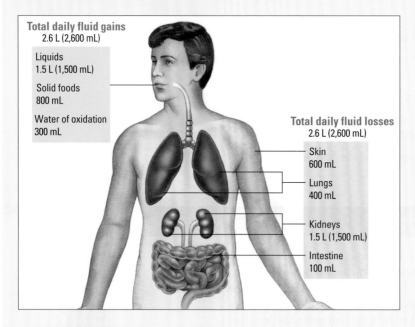

Total daily fluid gains
2.6 L (2,600 mL)

Liquids
1.5 L (1,500 mL)

Solid foods
800 mL

Water of oxidation
300 mL

Total daily fluid losses
2.6 L (2,600 mL)

Skin
600 mL

Lungs
400 mL

Kidneys
1.5 L (1,500 mL)

Intestine
100 mL

- *Interstitial fluid (ISF)* is found in the loose tissue around cells.
- *Extracellular fluid (ECF)* found in the spaces between cells, includes IVF and ISF.

Playing the percentages

ICF and ECF constitute about 40% and 20%, respectively, of an adult's total body weight. (See *Water weight*.)

Fluid forms and movement

Fluids in the body generally aren't found in pure forms. They're most commonly found in three different types of solutions: isotonic, hypotonic, and hypertonic.

Water weight

Water in the body exists in two major compartments that are separated by capillary walls and cell membranes. About two-thirds of the body's water is found within cells as intracellular fluid (ICF); the other third remains outside cells as extracellular fluid (ECF).

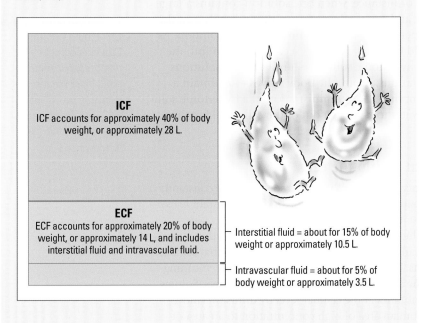

ICF
ICF accounts for approximately 40% of body weight, or approximately 28 L.

ECF
ECF accounts for approximately 20% of body weight, or approximately 14 L, and includes interstitial fluid and intravascular fluid.

Interstitial fluid = about for 15% of body weight or approximately 10.5 L.

Intravascular fluid = about for 5% of body weight or approximately 3.5 L.

No shifting needed

An isotonic solution has the same solute concentration as does another solution. For example, normal saline solution is considered isotonic because the concentration of sodium in the solution is nearly equal to the concentration of sodium in the blood. As a result, two equally concentrated fluids in adjacent compartments are already in balance so the fluid inside each compartment stays put; no imbalance means no net fluid shift. Cells won't shrink or swell because there's no gain or loss of water in the cell.

Go low for hypo...

A hypotonic solution has a lower solute concentration than does another solution. For instance, when one solution contains less sodium than does another solution, the first solution is hypotonic compared with the second. As a result, fluid from the first solution—the hypotonic solution—would shift into the second solution until the two solutions had equal concentrations. (Remember, the body constantly strives to maintain a state of balance, or equilibrium.) Administration of a hypotonic solution would cause water to move into the cells, making them swell.

...and high for hyper

A hypertonic solution has a higher solute concentration than does another solution. For instance, when one solution contains a large amount of sodium and a second solution contains hardly any, the first solution is hypertonic compared with the second solution. As a result, fluid would be drawn from the second solution into the first solution—the hypertonic solution—until the two solutions had equal concentrations. Again, the body strives to maintain a state of equilibrium; therefore, administration of a hypertonic solution would cause water to be drawn out of the cells, making them shrink.

Fluid movement within the cells

Just as the heart beats constantly, fluids and solutes move constantly within the body. That movement allows the body to maintain *homeostasis*, the constant state of balance the body seeks.

Solutes within the compartments of the body (intracellular, interstitial, and intravascular) move through the membranes that separate those compartments. The membranes are semipermeable, meaning that they allow some solutes to pass through but not others. Solutes move through membranes at the cellular level by diffusion (movement of particles from an area of high concentration to an area of lower concentration), active transport, or osmosis.

Upstream...

In *active transport*, solutes move from an area of lower concentration to an area of higher concentration. Think of active transport as swimming upstream. When a fish swims upstream, it has to expend energy.

...and against the current

The energy required for a solute to move against a concentration gradient comes from a substance called *adenosine triphosphate (ATP)*. Stored in all cells, ATP supplies energy for solute movement in and out of cells.

Some solutes, such as sodium and potassium, use ATP to move in and out of cells in a form of active transport called the *sodium-potassium pump*. Other solutes that require active transport to cross cell membranes include calcium ions, hydrogen ions, amino acids, and certain sugars.

Memory jogger

To help remember which fluid belongs to which compartment, keep in mind that *inter* means between (as in "interval") and *intra* means within or inside (as in "intravenous").

Just passing through

Osmosis refers to the passive movement of fluid across a membrane from an area of lower solute concentration and comparatively more fluid into an area of higher solute concentration and comparatively less fluid. Osmosis stops when enough fluid has moved through the membrane to equalize the solute concentration on both sides of the membrane.

In with the good

Water normally enters the body from the GI tract. Each day, the body obtains about 1.6 qt (1.5 L) of water from consumed liquids and approximately 26.6 oz (800 mL) more from solid foods, which may consist of up to 97% water. Oxidation of food in the body yields carbon dioxide (CO_2) and about 10 oz (300 mL) of water (water of oxidation).

Out with the bad

Water leaves the body through the skin (in perspiration), lungs (in expired air), GI tract (in stool), and urinary tract (in urine).

The major pipeline

The main route of water loss is urine excretion, which typically varies from 1 to 2.6 L daily. Water losses through the skin (600 mL) and lungs (400 mL) amount to 1 L daily but may increase markedly with strenuous exertion, which predisposes a person to dehydration.

Don't interrupt

In a healthy body, fluid gains match fluid losses to maintain proper physiologic functioning. However, interruption or dysfunction of one or both of the mechanisms that regulate fluid balance—thirst and the *countercurrent mechanism*—can lead to a fluid imbalance.

I'm parched!

Thirst—the conscious desire for water—is the primary regulator of fluid intake. When the body becomes dehydrated, ECF volume is reduced, causing an increase in sodium concentration and osmolarity.

When the sodium concentration reaches a level of about 2 mEq/L above normal, the neurons of the thirst center in the hypothalamus are stimulated. The brain then directs motor neurons to satisfy thirst, causing the person to drink enough fluid to restore ECF to normal levels.

What comes in must go out

Through the countercurrent mechanism, the kidneys regulate fluid output by modifying urine concentration—that is, by excreting urine of greater or lesser concentration, depending on fluid balance.

Thirst—the conscious desire for water—is the primary regulator of fluid intake.

Electrolyte balance

Electrolytes are substances that *dissociate* (break up) into electrically charged particles, called *ions*, when dissolved in water. Adequate amounts of each major electrolyte and a proper balance of electrolytes are required to maintain normal physiologic functioning.

All charged up

Ions may be positively charged (called *cations*) or negatively charged (called *anions*). Major cations include sodium, potassium, calcium, and magnesium. Major anions include chloride, bicarbonate (HCO_3^-), and phosphate.

Normally, the electrical charges of cations balance the electrical charges of anions, keeping body fluids electrically neutral. Blood plasma contains slightly more electrolytes than does ISF.

Shh! We're concentrating

Because ions are present in such low concentrations in body fluids, they're usually expressed in milliequivalents per liter (mEq/L). ICF and ECF cells are permeable to different substances; therefore, these compartments normally have different electrolyte compositions. (See *Electrolyte composition in ICF and ECF.*)

Electrolyte composition in ICF and ECF

This table shows the electrolyte compositions of intracellular fluid (ICF) and extracellular fluid (ECF).

Electrolyte	ICF	ECF
Sodium	10 mEq/L	136 to 146 mEq/L
Potassium	140 mEq/L	3.6 to 5 mEq/L
Calcium	10 mEq/L	4.5 to 5.8 mEq/L (ionized)
Magnesium	40 mEq/L	1.6 to 2.2 mEq/L
Chloride	4 mEq/L	96 to 106 mEq/L
Bicarbonate	10 mEq/L	24 to 28 mEq/L
Phosphate	100 mEq/L	1 to 1.5 mEq/L

A delicate balance

Electrolytes profoundly affect the body's water distribution, osmolarity, and acid-base balance. Numerous mechanisms within the body help maintain electrolyte balance. Dysfunction or interruption of any of these mechanisms can produce an electrolyte imbalance.

Here are the regulatory mechanisms for common electrolytes:

- The kidneys and a hormone called *aldosterone* are the chief sodium regulators. The small intestine absorbs sodium readily from food, and the skin and kidneys excrete sodium. (See *Osmotic regulation of sodium and water.*)

Now I get it!

Osmotic regulation of sodium and water

This flowchart illustrates two compensatory mechanisms used to restore sodium and water balance.

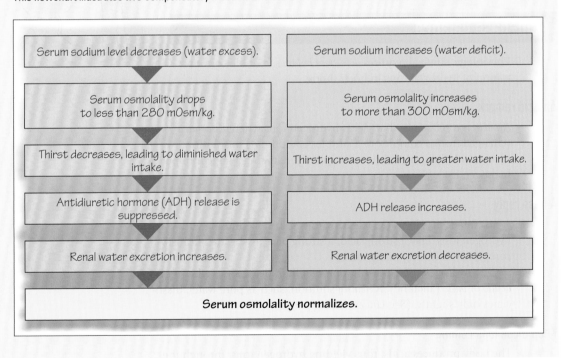

Serum sodium level decreases (water excess).	Serum sodium increases (water deficit).
Serum osmolality drops to less than 280 mOsm/kg.	Serum osmolality increases to more than 300 mOsm/kg.
Thirst decreases, leading to diminished water intake.	Thirst increases, leading to greater water intake.
Antidiuretic hormone (ADH) release is suppressed.	ADH release increases.
Renal water excretion increases.	Renal water excretion decreases.

Serum osmolality normalizes.

- The kidneys also regulate potassium through aldosterone action. Most potassium is absorbed from food in the GI tract; normally, the amount excreted in urine equals dietary potassium intake.
- Calcium in the blood is typically in equilibrium with calcium salts in bone. Parathyroid hormone (PTH) is the main regulator of calcium, controlling both calcium uptake from the GI tract and calcium excretion by the kidneys.
- Magnesium is governed by aldosterone, which controls renal magnesium reabsorption. Absorbed from the GI tract, magnesium is excreted in urine, breast milk, and saliva.
- The kidneys also regulate chloride. Chloride ions move in conjunction with sodium ions.
- The kidneys regulate bicarbonate, excreting, absorbing, or forming it. Bicarbonate, in turn, plays a vital part in acid-base balance.
- The kidneys regulate phosphate. Absorbed from food, phosphate is incorporated with calcium in bone. PTH governs calcium and phosphate levels.

You're an electrolyte superstar—helping to regulate potassium, chloride, bicarbonate, and phosphate.

Acid–base balance

Physiologic survival requires *acid-base balance*, a stable concentration of hydrogen ions in body fluids.

An acid remark

An *acid* is a substance that yields hydrogen ions when *dissociated* (changed from a complex to a simpler compound) in solution. A strong acid dissociates almost completely, releasing a large number of hydrogen ions.

A base reply

A *base* dissociates in water, releasing ions that can combine with hydrogen ions. Like a strong acid, a strong base dissociates almost completely, releasing many ions.

The hydrogen ion concentration of a fluid determines whether it's *acidic* or *basic* (alkaline). A *neutral* solution, such as pure water, dissociates only slightly. (See *Understanding pH*.)

Keep those ions coming

The body produces acids, thus yielding hydrogen ions, through the following mechanisms:
- Protein catabolism yields nonvolatile acids, such as sulfuric, phosphoric, and uric acids.
- Fat oxidation produces acid ketone bodies.

Understanding pH

Hydrogen (H) ion concentration is commonly expressed as pH, which indicates the degree of acidity or alkalinity of a solution:
- A pH of 7 indicates neutrality or equal amounts of H and hydroxyl (OH⁻) ions.
- An acidic solution contains more H ions than OH⁻ ions; its pH is less than 7.
- An alkaline solution contains fewer H ions than OH⁻ ions; its pH exceeds 7.
  Overall, as H ion concentration increases, pH goes down.

1 means 10
Because pH is an exponential expression, a change of one pH unit reflects a 10-fold difference in actual H ion concentration. For instance, a solution with a pH of 7 has 10 times more H ions than does a solution with a pH of 8.

- Anaerobic glucose catabolism produces lactic acid.
- Intracellular metabolism yields CO_2 as a by-product; CO_2 dissolves in body fluids to form carbonic acid (H_2CO_3).

Balancing buffers

Normally, even with the production of these acids, the body's pH control mechanism is so effective that blood pH stays within a narrow range: 7.35 to 7.45. This acid-base balance is maintained by buffer systems and the lungs and kidneys, which neutralize and eliminate acids as rapidly as they're formed.

Dysfunction or interruption of a buffer system or other governing mechanism can cause an acid-base imbalance. *Acidosis* occurs when the hydrogen ion concentration increases above normal. *Alkalosis* occurs when the hydrogen ion concentration falls below normal. Acidosis or alkalosis can occur in the body as a result of respiratory or metabolic disorders, or a combination of both. (See *Understanding respiratory and metabolic alkalosis and acidosis*, pages 258 to 261.)

(Text continues on page 262)

Now I get it!

Understanding respiratory and metabolic alkalosis and acidosis

What happens in respiratory alkalosis
Step 1
When pulmonary ventilation increases above the amount needed to maintain normal carbon dioxide (CO_2) levels, excessive amounts of CO_2 are exhaled. This causes hypocapnia (a fall in partial pressure of arterial carbon dioxide [$Paco_2$]), which leads to a reduction in carbonic acid (H_2CO_3) production, a loss of hydrogen (H) ions and bicarbonate (HCO_3^-) ions, and a subsequent rise in pH. Look for a pH level above 7.45, a $Paco_2$ level below 35 mm Hg, and an HCO_3^- level below 22 mEq/L (as shown at right).

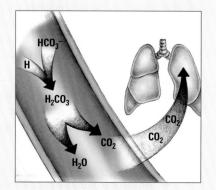

Step 2
In defense against the rising pH, H ions are pulled out of the cells and into the blood in exchange for potassium (K) ions. The H ions entering the blood combine with HCO_3^- ions to form H_2CO_3, which lowers the pH. Look for a further decrease in HCO_3^- levels, a fall in pH, and a fall in serum K levels (hypokalemia).

Step 3
Hypocapnia stimulates the carotid and aortic bodies and the medulla, which causes an increase in heart rate without an increase in blood pressure (as shown at right). Look for angina, electrocardiogram changes, restlessness, and anxiety.

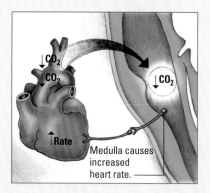

Step 4
Simultaneously, hypocapnia produces cerebral vasoconstriction, which prompts a reduction in cerebral blood flow. Hypocapnia also overexcites the medulla, pons, and other parts of the autonomic nervous system. Look for increasing anxiety, diaphoresis, dyspnea, alternating periods of apnea and hyperventilation, dizziness, and tingling in the fingers or toes.

Step 5
When hypocapnia lasts more than 6 hours, the kidneys increase secretion of HCO_3^- and reduce excretion of H (as shown at right). Periods of apnea may result if the pH remains high and the $Paco_2$ remains low. Look for slowing of the respiratory rate, hypoventilation, and Cheyne-Stokes respirations.

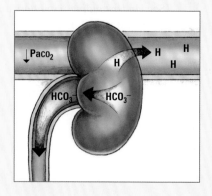

Step 6
Continued low $Paco_2$ increases cerebral and peripheral hypoxia from vasoconstriction. Severe alkalosis inhibits calcium (Ca) ionization which, in turn, causes increased nerve excitability and muscle contractions. Eventually, the alkalosis overwhelms the central nervous system (CNS) and the heart. Look for decreasing level of consciousness (LOC), hyperreflexia, carpopedal spasm, tetany, arrhythmias, seizures, and coma.

Understanding respiratory and metabolic alkalosis and acidosis *(continued)*

What happens in respiratory acidosis
Step 1
When pulmonary ventilation decreases, retained CO_2 combines with water (H_2O) to form H_2CO_3 in larger-than-normal amounts. The H_2CO_3 dissociates to release free H and HCO_3^- ions. The excessive H_2CO_3 causes a drop in pH. Look for a $Paco_2$ level above 45 mm Hg and a pH level below 7.35.

Step 2
As the pH level falls, 2,3-diphosphoglycerate (2,3-DPG) increases in the red blood cells and causes a change in hemoglobin (Hb) that makes the Hb release oxygen (O_2). The altered Hb, now strongly alkaline, picks up H ions and CO_2, thus eliminating some of the free H ions and excess CO_2 (as shown at right). Look for decreased arterial oxygen saturation.

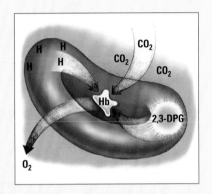

Step 3
Whenever $Paco_2$ increases, CO_2 builds up in all tissues and fluids, including cerebrospinal fluid and the respiratory center in the medulla. The CO_2 reacts with H_2O to form H_2CO_3, which then breaks into free H and HCO_3^- ions. The increased amount of CO_2 and free H ions stimulate the respiratory center to increase the respiratory rate. An increased respiratory rate expels more CO_2 and helps to reduce the CO_2 level in the blood and other tissues. Look for rapid, shallow respirations and a decreasing $Paco_2$.

Step 4
Eventually, CO_2 and H ions cause cerebral blood vessels to dilate, which increases blood flow to the brain. That increased flow can cause cerebral edema and depress CNS activity. Look for headache, confusion, lethargy, nausea, or vomiting. As respiratory mechanisms fail, the increasing $Paco_2$ stimulates the kidneys to retain HCO_3^- and sodium (Na) ions and to excrete H ions, some of which are excreted in the form of ammonium (NH_4). The additional HCO_3^- and Na combine to form extra sodium bicarbonate ($NaHCO_3^-$), which is then able to buffer more free H ions (as shown at right). Look for increased acid content in the urine, increasing serum pH and HCO_3^- levels, and shallow, depressed respirations.

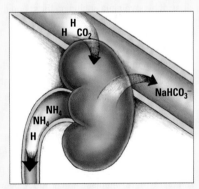

Step 5
As the concentration of H ions overwhelms the body's compensatory mechanisms, the H ions move into the cells and K ions move out. A concurrent lack of O_2 causes an increase in the anaerobic production of lactic acid, which further skews the acid-base balance and critically depresses neurologic and cardiac functions. Look for hyperkalemia, arrhythmias, increased $Paco_2$, decreased partial pressure of arterial oxygen, decreased pH, and decreased LOC.

(continued)

Understanding respiratory and metabolic alkalosis and acidosis *(continued)*

What happens in metabolic alkalosis
Step 1
As HCO_3^- ions start to accumulate in the body, chemical buffers (in extra-cellular fluid [ECF] and cells) bind with the ions. No signs are detectable at this stage.

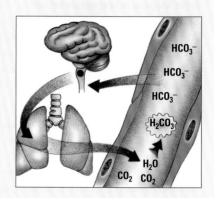

Step 2
Excess HCO_3^- ions that don't bind with chemical buffers elevate serum pH levels, which, in turn, depress chemoreceptors in the medulla. Depression of those chemoreceptors causes a decrease in respiratory rate, which increases the $Paco_2$. The additional CO_2 combines with H_2O to form H_2CO_3 (as shown at right). Note: Lowered O_2 levels limit respiratory compensation. Look for a serum pH level above 7.45, an HCO_3^- level above 26 mEq/L, a rising $Paco_2$, and slow, shallow respirations.

Step 3
When the HCO_3^- level exceeds 28 mEq/L, the renal glomeruli can no longer reabsorb excess HCO_3^-. That excess HCO_3^- is excreted in the urine; H ions are retained. Look for alkaline urine and pH and HCO_3^- levels that slowly return to normal.

Step 4
To maintain electrochemical balance, the kidneys excrete excess Na ions, H_2O, and HCO_3^- (as shown at right). Look for polyuria initially and then signs of hypovolemia, including thirst and dry mucous membranes.

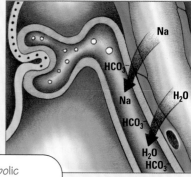

Step 5
Lowered H ion levels in the ECF cause the ions to diffuse out of the cells. To maintain the balance of charge across the cell membrane, extracellular K ions move into the cells. Look for signs of hypokalemia: anorexia, muscle weakness, loss of reflexes, and others.

Step 6
As H ion levels decline, Ca ionization decreases. That decrease in ionization makes nerve cells more permeable to Na ions. Na ions moving into nerve cells stimulate neural impulses and produce overexcitability of the peripheral nervous system and CNS. Look for tetany, belligerence, irritability, disorientation, and seizures.

Metabolic alkalosis ultimately progresses to tetany, belligerence, irritability, disorientation, and seizures.

Understanding respiratory and metabolic alkalosis and acidosis *(continued)*

What happens in metabolic acidosis

Step 1

As H ions start to accumulate in the body, chemical buffers (plasma HCO_3^- and proteins) in the cells and ECF bind with them (as shown at right). No signs are detectable at this stage.

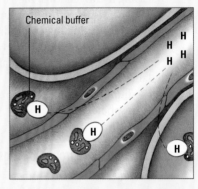

Step 2

Excess H ions (which can't bind with the buffers) decrease the pH and stimulate chemoreceptors in the medulla to increase the respiratory rate. The increased respiratory rate lowers the $Paco_2$, which allows more H ions to bind with HCO_3^- ions. Respiratory compensation occurs within minutes, but isn't sufficient to correct the imbalance (see middle illustration). Look for a pH level below 7.35, an HCO_3^- level below 22 mEq/L, a decreasing $Paco_2$ level, and rapid, deeper respirations.

Step 3

Healthy kidneys try to compensate for the acidosis by secreting excess H ions into the renal tubules. Those ions are buffered by phosphate or ammonia and then are excreted into the urine in the form of a weak acid. Look for acidic urine.

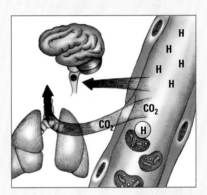

Step 4

Each time an H ion is secreted into the renal tubules, an Na ion and an HCO_3^- ion are absorbed from the tubules and returned to the blood. Look for pH and HCO_3^- levels that slowly return to normal.

Step 5

Excess H ions in the ECF diffuse into cells. To maintain the balance of the charge across the membrane, the cells release K ions into the blood (as shown at right). Look for signs of hyperkalemia, including colic and diarrhea, weakness or flaccid paralysis, tingling and numbness in the extremities, bradycardia, a tall T wave, a prolonged PR interval, and a wide QRS complex.

Step 6

Excess H ions alter the normal balance of K, Na, and Ca ions, leading to reduced excitability of nerve cells. Look for signs and symptoms of progressive CNS depression, including lethargy, dull headache, confusion, stupor, and coma.

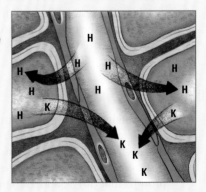

Buffer systems

Buffers are substances that prevent changes in the pH by removing or releasing hydrogen ions. *Buffer systems* reduce the effect of an abrupt change in hydrogen ion concentration by converting a strong acid or base (which normally would dissociate completely) into a weak acid or base (which releases fewer hydrogen ions).

Buffer systems that help maintain acid-base balance include:

- sodium bicarbonate–carbonic acid
- phosphate
- protein.

One from the kidneys, one from the lungs

The *sodium bicarbonate–carbonic acid* buffer system is the major buffer in ECF. Sodium bicarbonate concentration is regulated by the kidneys, and carbonic acid concentration is regulated by the lungs. Both components of this buffer are replenished continually. As a result of the buffering action, the strong base (sodium hydroxide) is replaced by sodium bicarbonate and water (H_2O). Sodium hydroxide dissociates almost completely and releases large amounts of hydroxyl. If a strong acid is added, the opposite occurs.

Finesse with phosphate

A *phosphate* buffer system works by regulating the pH of fluids as they pass through the kidneys. It's also important in ECF.

Absorbing ions

Protein buffers can exist in the form of acids or alkaline salts. In the *protein* buffer system, intracellular proteins absorb hydrogen (H^+) ions generated by the body's metabolic processes and may release excess hydrogen as needed.

Lungs

The protein buffer system changes the pH of the blood in 3 minutes or less by changing the respiratory rate. A decreased respiratory rate decreases the exchange and release of CO_2; there also is less hydrogen, and the pH rises. Present in all acids, hydrogen ions are protons that can be added to or removed from a solution to change the pH.

Respiration plays a crucial role in controlling pH. The lungs excrete CO_2 and regulate the carbonic acid content of the blood. Carbonic acid is derived from the CO_2 and water that are released as by-products of cellular metabolic activity.

Stick to the formula

CO_2 is soluble in blood plasma. Some of the dissolved gas reacts with water to form carbonic acid, a weak acid that partially breaks apart to form hydrogen and bicarbonate ions. These three substances are in equilibrium, as reflected in the following formula:

$$CO_2 + H_2O \leftrightarrow H_2CO_3 \leftrightarrow H^+ + HCO_3^-.$$

CO_2 dissolved in plasma is in equilibrium with CO_2 in the lung alveoli (expressed as a partial pressure [P_{CO_2}]). Thus, an equilibrium exists between alveolar P_{CO_2} and the various forms of CO_2 present in the plasma, as expressed by the following formula:

$$P_{CO_2} \leftrightarrow CO_2 + H_2O \leftrightarrow H_2CO_3 \leftrightarrow H^+ + HCO_3^-.$$

Don't hold your breath!

A change in the rate or depth of respirations can alter the CO_2 content of alveolar air and the alveolar P_{CO_2}. A change in alveolar P_{CO_2} produces a corresponding change in the amount of carbonic acid formed by dissolved CO_2. In turn, these changes stimulate the respiratory center to modify respiratory rate and depth.

An increase in alveolar P_{CO_2} raises the blood concentration of CO_2 and carbonic acid. This, in turn, stimulates the respiratory center to increase respiratory rate and depth. As a result, alveolar P_{CO_2} decreases, which leads to a corresponding drop in the carbonic acid and CO_2 concentrations in blood. (See *How respiratory mechanisms affect blood pH*, page 264.)

A decrease in respiratory rate and depth has the reverse effect; it raises alveolar P_{CO_2} which, in turn, triggers an increase in the blood's CO_2 and carbonic acid concentrations.

Kidneys

In addition to excreting various acid waste products, the kidneys help manage acid-base balance by regulating the blood's bicarbonate concentration. They do so by permitting bicarbonate reabsorption from tubular filtrate and by forming additional bicarbonate to replace that used in buffering acids.

How respiratory mechanisms affect blood pH

A decrease in blood pH stimulates the respiratory center, causing *hyperventilation*. As a result, carbon dioxide (CO_2) levels decrease and, therefore, less carbonic acid and fewer hydrogen ions remain in the blood. Consequently, blood pH increases, possibly reaching a normal level.

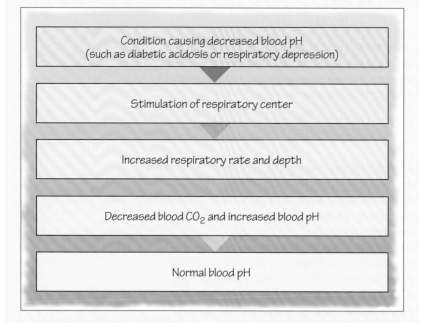

Condition causing decreased blood pH
(such as diabetic acidosis or respiratory depression)

Stimulation of respiratory center

Increased respiratory rate and depth

Decreased blood CO_2 and increased blood pH

Normal blood pH

A rise in carbon dioxide or a decrease in pH in the blood causes hyperventilation. That means we have to work harder!

Renal tubular ion secretion

Recovery and formation of bicarbonate in the kidneys depend on hydrogen ion secretion by the renal tubules in exchange for sodium ions. Sodium ions are then simultaneously reabsorbed into the circulation from the tubular filtrate.

Influential enzyme

The enzyme *carbonic anhydrase* influences tubular epithelial cells to form carbonic acid from CO_2 and water. Carbonic acid quickly dissociates into hydrogen and bicarbonate ions. Hydrogen ions enter the tubular filtrate in exchange for sodium ions; bicarbonate ions enter the bloodstream along with the sodium ions that have been absorbed from the filtrate. Bicarbonate is then reabsorbed from the tubular filtrate.

Ions hanging out together

Each hydrogen ion secreted into the tubular filtrate joins with a bicarbonate ion to form carbonic acid, which rapidly dissociates into CO_2 and water. The CO_2 diffuses into the tubular epithelial cells, where it can combine with more water and lead to the formation of more carbonic acid.

Bicarbonate reabsorption

The remaining water molecule in the tubular filtrate is eliminated in the urine. As each hydrogen ion enters the tubular filtrate to combine with a bicarbonate ion, a bicarbonate ion in the tubular epithelial cells diffuses into the circulation. This process is termed bicarbonate reabsorption. (However, the bicarbonate ion that enters the circulation isn't the same one as in the tubular filtrate.)

Formation of ammonia and phosphate salts

To form more bicarbonate, the kidneys must secrete additional hydrogen ions in exchange for sodium ions. For the renal tubules to continue secreting hydrogen ions, the excess ions must combine with other substances in the filtrate and be excreted. Excess hydrogen ions in the filtrate may combine with *ammonia* (NH_3), which is produced by the renal tubules, or with phosphate salts present in the tubular filtrate.

More ions on the move

After diffusing into the filtrate, ammonia joins with the secreted hydrogen ions, forming ammonium ions. These ions are excreted in the urine with chloride and other anions; each secreted ammonia molecule eliminates one hydrogen ion in the filtrate.

At the same time, sodium ions that have been absorbed from the filtrate and exchanged for hydrogen ions enter the circulation, as does the bicarbonate formed in the tubular epithelial cells. Some secreted hydrogen ions combine with a *disodium hydrogen phosphate* (Na_2HPO_4) in the tubular filtrate. Each of the secreted hydrogen ions that joins with the disodium salt changes to the monosodium salt sodium dihydrogen phosphate (NaH_2PO_4). The sodium ion released in this reaction is absorbed into the circulation along with a newly formed bicarbonate ion.

Factors affecting bicarbonate formation

The rate of bicarbonate formation by renal tubular epithelial cells is affected by two factors:

- the amount of dissolved CO_2 in the plasma
- the potassium content of the tubular cells.

If the plasma CO_2 level rises, renal tubular cells form more bicarbonate. Increased plasma CO_2 encourages greater carbonic acid formation by the renal tubular cells.

Chain reaction

Partial dissociation of carbonic acid results in more hydrogen ions for excretion into the tubular filtrate and additional bicarbonate ions for entry into the circulation. This, in turn, increases the plasma bicarbonate level and reduces the plasma level of dissolved CO_2 toward normal. If the plasma CO_2 level decreases, renal tubular cells form less carbonic acid.

Because fewer hydrogen ions are formed and excreted, fewer bicarbonate ions enter the circulation. The plasma bicarbonate level then falls accordingly.

> Potassium secretion by tubular epithelial cells decreases and hydrogen ion secretion rises in patients with potassium depletion from vomiting or diarrhea.

Special K

The potassium content of renal tubular cells also helps regulate plasma bicarbonate concentration by affecting the rate at which the renal tubules secrete hydrogen ions. Tubular cell potassium content and hydrogen ion secretion are interrelated; potassium and hydrogen ions are secreted at rates that vary inversely. Hydrogen ion secretion increases if tubular secretion of potassium ions falls; hydrogen secretion declines if tubular secretion of potassium ions increases.

For each hydrogen ion secreted into the tubular filtrate, an additional bicarbonate ion enters the blood plasma. Consequently, increased tubular secretion of hydrogen ions leads to a rise in the plasma bicarbonate content.

Depletion of body potassium causes more bicarbonate to enter the circulation; the plasma bicarbonate level then rises above normal. When the body contains excess potassium, the tubules excrete more potassium. As a result, fewer hydrogen ions are secreted, less bicarbonate forms, and the plasma bicarbonate concentration decreases.

Quick quiz

1. The nurse knows that a solution with a pH of less than 7 is considered to be which type of solution?
   A. Acidic
   B. Alkaline
   C. Solute
   D. Hypotonic

Answer: A. A solution with a pH of less than 7 contains more hydrogen ions than hydroxyl ions and is considered acidic.

2. The nurse is caring for a patient with respiratory alkalosis and knows that the way the body compensates for this change is by developing which of the following?
   A. Metabolic alkalosis
   B. Respiratory acidosis
   C. Metabolic acidosis
   D. A phosphate buffer system

Answer: C. The body compensates for chronic respiratory alkalosis by developing metabolic acidosis.

3. The nurse knows that there are two factors that affect the rate of bicarbonate formation by renal tubular epithelial cells. These factors are:
   A. amount of aldosterone in the system and urine production.
   B. amount of dissolved CO_2 in the plasma and the potassium content of the tubular cells.
   C. amount of dissolved potassium in the plasma and the CO_2 content of the tubular cells.
   D. amount of ammonia produced by the renal tubules and the phosphate salts present in the tubular filtrate.

Answer: B. The rate of bicarbonate formation by renal tubular epithelial cells is affected by the amount of dissolved CO_2 in the plasma and the potassium content of tubular cells.

Scoring

★★★ If you answered all three questions correctly, give yourself a pat on the back! You have all your knowledge in the proper balance.

★★ If you answered two questions correctly, good for you! You're benefiting from all the right buffer systems.

★ If you answered only one question correctly, check out this chapter again. It may take time to find your equilibrium.

Just for fun!

Unscramble the words on the left to discover the names of buffer systems that help maintain acid-base balance. Then draw lines linking each system to its particular characteristics.

DOSIUM
ACABBIETORN-RANBOCCI CADI

_ _ _ _ _ _ _

_ _ _ _ _ _ _ _ _ _ _

_ _ _ _ _ _ _ _ _

_ _ _ _

HEATHPOPS

_ _ _ _ _ _ _ _ _

POINTER

_ _ _ _ _ _ _

A. The major buffer in ECF
B. Can exist in the form of acids or alkaline salts
C. Regulates the pH of fluids as they pass through the kidneys
D. Can alter breathing rate to change blood pH
E. Uses the kidneys to regulate sodium bicarbonate concentration and the lungs to regulate carbonic acid
F. Has sodium dihydrogen phosphate as its acidic component and sodium monohydrogen phosphate as its alkaline component
G. Replaces the strong base (sodium hydroxide) with sodium bicarbonate and water; sodium hydroxide dissociates and releases large amounts of hydroxyl
H. Uses intracellular proteins to absorb hydrogen ions

Answer: SODIUM BICARBONATE–CARBONIC ACID: A, E, G; PHOSPHATE: C, F; PROTEIN: B, D, H

Selected References

Chabner, D. (2016). *The language of medicine* (11th ed.). St. Louis, MO: Elsevier.

Emmitt, M. (2014). *Simple and mixed acid base disorders. Up To Date.* Retrieved from http://www.uptodate.com/contents/simple-and-mixed-acid-base-disorders?source=search_result&search=acid+base+balance&selectedTitle=1%7E48

Grossman, S., & Porth, C. M. (2013). *Porth's pathophysiology: Concepts of altered health states* (9th ed.). Philadelphia, PA: Lippincott.

Hinkle, J., & Cheever, K. (2013). *Brunner and Suddarth's textbook of medical surgical nursing* (13th ed.). Philadelphia, PA: Lippincott.

Hogan, J., & Goldfarb, S. (2014). *Regulation of calcium and phosphate balance. Up To Date.* Retrieved from http://www.uptodate.com/contents/regulation-of-calcium-and-phosphate-balance?source=search_result&search=serum+calcium&selectedTitle=8%7E150

Mount, D. B. (2014). *Causes and evaluation of hyperkalemia in adults.* Retrieved from http://www.uptodate.com/contents/causes-and-evaluation-of-hyperkalemia-in-adults?source=search_result&search=potassium+regulation&selectedTitle=1%7E150

Sterns, R. H. (2014). *General principles of disorders of water balance (hyponatremia and hypernatremia) and sodium balance (hypovolemia and edema).* Retrieved from http://www.uptodate.com/contents/general-principles-of-disorders-of-water-balance-hyponatremia-and-hypernatremia-and-sodium-balance-hypovolemia-and-edema?source=search_result&search=osmotic+regulation+of+sodium+and+water&selectedTitle=1%7E150

Theodore, A. C. (2015). *Arterial blood gases. Up To Date.* Retrieved from http://www.uptodate.com/contents/arterial-blood-gases?source=search_result&search=blood+gas+interpretation&selectedTitle=1%7E150

Chapter 16

Reproductive system

Just the facts

In this chapter, you'll learn:

♦ anatomic structure and functions of the male and female reproductive systems

♦ male hormone production and its effects on sexual development

♦ female hormone production and its effects on menstruation

♦ anatomic structure and functions of the female breast.

A look at the reproductive systems

Anatomically, the main distinction between the male and female is the presence of conspicuous external genitalia in the male versus the internal (within the pelvic cavity) location of the major reproductive organs in the female.

> Here's the main difference—the majority of the male organs for reproduction are external, but most female reproductive organs are inside the pelvic cavity.

Male reproductive system

The male reproductive system consists of the organs that produce, transfer, and introduce mature sperm into the female reproductive tract, where fertilization occurs. (See *Structures of the male reproductive system.*)

Structures of the male reproductive system

The male reproductive system consists of the penis, the scrotum and its contents, the prostate gland, and the inguinal structures. These structures are illustrated below. (Merck Manual, Gray's Anatomy, Netter's Atlas)

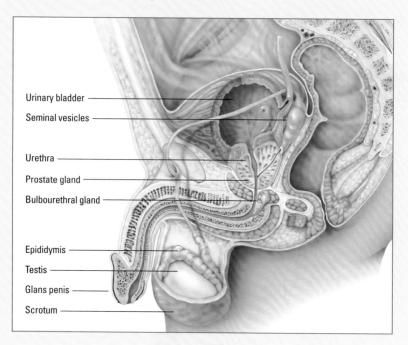

Extra work

In addition to forming male sex cells (spermatogenesis), the male reproductive system plays a role in the secretion of male sex hormones, primarily testosterone.

Penis

An organ of copulation and means of evacuation, the *penis* deposits sperm in the female reproductive tract and acts as the terminal duct for the urinary tract. It consists of an attached root, a free shaft, and an enlarged tip, or glans penis.

Internally, the cylinder-shaped penile shaft consists of three columns of *erectile tissue* bound together by *heavy fibrous tissue and smooth muscle*. Two lateral columns of cavernous erectile tissue (*corpus cavernosum*) form the majority of the penile shaft. These columns meet the *corpus spongiosum*, which surrounds the urethra from the penile bulb at its proximal root until it terminates at the external urethral os at the glans penis (Waugh & Grant, 2014, pp. 461–463; Wilkins, Cross, Megson, & Meredith, 2011, pp. 624–625).

The *glans penis*, at the distal end of the shaft, is a cone-shaped structure formed from the corpus spongiosum. Its lateral margin forms a ridge of tissue known as the *corona*. Just posterior to the glans, the skin folds to form the foreskin or prepuce. The prepuce can be removed surgically with a procedure called circumcision (Shier, Butler, & Lewis, 2015, pp. 524–526). The glans penis is highly sensitive to sexual stimulation (Faiz, Blackburn, & Moffat, 2011, pp. 72–73; Waugh & Grant, 2014, pp. 461–463). The *urethral meatus* opens through the glans to allow urination and ejaculation.

In a different vein

The penis receives blood through the *internal pudendal artery*. Blood then flows into the corpora cavernosa through the penile artery. Venous blood returns through the *internal iliac vein* to the *vena cava*.

Scrotum

The penis meets the *scrotum*, or scrotal sac, at the penoscrotal junction. Located posterior to the penis and anterior to the anus, the scrotum is an extra-abdominal pouch that consists of a thin layer of skin overlying a layer of superficial fascia and smooth muscle.

Internally, a *septum* divides the scrotum into two sacs, which each contain a *testis* (Marieb & Hoehn, 2014, pp. 1093–1094; Shier et al., 2015, pp. 524–525).

Loads of nodes

Lymph nodes from the penis, scrotal surface, and anus drain into the *inguinal* lymph nodes. Lymph nodes from the testes drain into the lateral aortic and preaortic lymph nodes in the abdomen.

An important function of the scrotum is to keep the testes cooler than the rest of the body by approximately 3°C.

Testes

The testes are the reproductive glands of the male. There are usually two testes approximately 4 × 2.5 cm in size enclosed in the scrotal sac by three layers of serous fibrous tissue: the *tunica vaginalis*, the *tunica albuginea* and the *tunica vasculosa* (Marieb & Hoehn, 2014, pp. 1094–1095). Extensions of the tunica albuginea divide the testes glands into 200 to 300 *lobules*. Each lobule contains one to four *seminiferous tubules*, small tubes in which spermatogenesis takes place (Waugh & Grant, 2014, pp. 459–461).

Climate control

Spermatozoa development requires a temperature lower than that of the rest of the body. The *dartos muscle*, a smooth muscle in the superficial fascia, causes scrotal skin to wrinkle, which helps to regulate temperature. The *cremaster muscle*, rising from the internal oblique muscle, helps to govern temperature by elevating the testes.

Duct system

The male reproductive *duct system*, consisting of the epididymis and vas deferens, conveys sperm from the testes to the ejaculatory ducts near the bladder and terminates at the external urethral orifice (Marieb & Hoehn, 2014, pp. 1098–1099).

Storage

Once formed, spermatozoa leave the testis via a tubular network to the *epididymis* where they are stored until ejaculation. The *epididymides* are coiled tubes approximately 18 to 20 in length that emerge at the superior aspect of each testis and lie along the posterior border. During ejaculation, smooth muscle in the epididymis contracts, ejecting spermatozoa into the *vas deferens*.

Descending and merging

The *vas deferens* (also called the *ductus deferens*) is a muscular tube that conveys sperm from the epididymis to the ejaculatory duct. This system leads from the testes to the abdominal cavity, where it extends upward through the *inguinal canal*, arches over the urethra, and descends behind the bladder. Its enlarged portion, called the *ampulla*, merges with the duct of the seminal vesicle to form the short ejaculatory duct. After passing through the prostate gland, the vas deferens joins with the urethra (Marieb & Hoehn, 2014, pp. 1096–1099).

The vas deferens is one of the structures in the *spermatic cord*, a connective tissue sheath that travels from the testes through the inguinal canal, exiting the scrotum through the external inguinal ring and entering the abdominal cavity through the internal inguinal ring. The inguinal canal lies between the two rings. The spermatic cord encases autonomic nerve fibers, blood vessels, lymph vessels, and the vas deferens (Shier et al., 2015, pp. 520–524; Waugh & Grant, 2014, pp. 459–461).

Tubular exit

A small tube leading from the floor of the bladder to the exterior, the *urethra* consists of three parts:

- *prostatic urethra* (surrounded by the prostate gland), which drains the bladder
- *membranous urethra*, which passes through the urogenital diaphragm
- *spongy urethra*, which makes up about 75% of the entire urethra (Marieb & Hoehn, 2014).

Accessory reproductive glands

The accessory reproductive glands, which produce most of the semen, include the *seminal vesicles, bulbourethral glands (Cowper's glands)*, and the *prostate gland*. The seminal vesicles are 5-cm paired sacs attached to the vas deferens at the base of the urinary bladder. The two bulbourethral glands are approximately one centimeter and are located inferior to the prostate gland.

Size of a walnut

The prostate is a ring-shaped gland approximately 4 × 3 cm thick and lies under the bladder and surrounds the proximal urethra. The prostate is divided into several lobes: the left and right lateral lobes make up the main mass of the glandular tissue, the median lobe is a cone-shaped area that lies between the two ejaculatory ducts and the urethra, and the anterior lobe is composed of fibromuscular tissue and contains no glandular tissue (Marieb & Hoehn, 2014, pp. 1099–2000).

Improving the odds

The seminal glands (seminal vesicles) are 5- to 7-cm hollow glands made up of smooth muscle surrounding a coiled mucosa of secretory pseudostratified columnar epithelium that secretes and stores seminal fluid. Seminal fluid is a yellow, viscous, alkaline substance made up of fructose, citric acid, prostaglandins, and enzymes. This fluid is what fluoresces under UV light during investigation of sexual attacks. In addition, seminal fluid may enhance sperm motility and fertilization. It makes up 60% to 70% of the volume of ejaculate. This gland meets the vas deferens at the ejaculatory duct, and seminal fluid combines with sperm prior to combining with prostatic fluid (Marieb & Hoehn, 2014, pp. 1099–1100).

The prostate continuously secretes prostatic fluid, a thin, milky, acidic fluid. During ejaculation, contraction of prostatic smooth muscle forces this fluid into the prostatic urethra. Prostatic fluid plays a part in activating sperm and makes up approximately one-third of the volume of the ejaculate. The fluid is made up of citrate, enzymes, and prostate-specific antigen (Marieb & Hoehn, 2014, pp. 1099–2000; Waugh & Grant, 2014, pp. 459–461).

Slightly alkaline

Semen is a viscous, white secretion with a slightly alkaline pH (7.2 to 8); it consists of spermatozoa and accessory gland secretions. The seminal vesicles produce roughly 60% to 70% of the fluid portion of the semen, while the prostate gland produces about 30%. A viscid fluid secreted by the bulbourethral glands to neutralize trace urine and lubricate the urethra also becomes part of the semen (Marieb & Hoehn, 2014, pp. 1099–2000).

> Seminal fluid enhances sperm motility and may increase the chances for conception by neutralizing the acidity of the man's urethra and the woman's vagina.

Spermatogenesis

Sperm formation, or *spermatogenesis*, begins when a male reaches *puberty* and normally continues throughout life.

Divide and conquer

Spermatogenesis occurs in four stages over 64 to 72 days:
1. Mitosis: The primary germinal epithelial cells, called *spermatogonia*, grow and develop into primary *spermatocytes*. Both spermatogonia and primary spermatocytes contain 46 chromosomes, consisting of 44 *autosomes* and the two sex chromosomes, X and Y.

Memory jogger

To remember the meaning of spermatogenesis, keep in mind that genesis means "beginning" or "new." Therefore, spermatogenesis means beginning of new sperm.

2. Meiosis I: Primary spermatocytes divide to form secondary spermatocytes. No new chromosomes are formed in this stage; the pairs only divide. Each secondary spermatocyte contains one-half the number of autosomes, 22. One secondary spermatocyte contains an X chromosome; the other, a Y chromosome.

3. Meiosis II: Each secondary spermatocyte divides again to form *spermatids* (also called *spermatoblasts*).

4. Spermiogenesis: The spermatids undergo a series of structural changes that transform them into mature *spermatozoa*, or sperm. Each spermatozoa has a head, neck, midsection, and tail. The head contains the *nucleus*; the tail contains a large amount of *adenosine triphosphate*, which provides energy for sperm *motility* (Marieb & Hoehn, 2014, pp. 1101–1107).

Queuing up

New sperms pass from the seminiferous tubules through the tubular network of the *rete testis* to the efferent ductules and into the epididymis for maturation and storage (Marieb & Hoehn, 2014, pp. 1093–1095; Smith & Walker, 2014).

> I'm newly mature and ready for emission.

Keeps for weeks

Sperm cells retain their potency in storage for many weeks. After ejaculation, sperm can survive 3 to 5 days in the female reproductive tract.

Male hormonal control and sexual development

Androgens (male sex hormones) are primarily produced in the testes, though small amounts are made in the adrenal glands. They're responsible for the development of male sex organs and secondary sex characteristics. Important male reproductive hormones include:

- GnRH (gonadotropin-releasing hormone)
- gonadotropins
- luteinizing hormone (LH) and follicle-stimulating hormone (FSH)
- testosterone
- inhibin (Smith & Walker, 2014).

> Number and motility affect fertility. A sperm count less than 15 million per milliliter of ejaculated semen or progressive motility less than 32% may result in decreased fertility (Cooper & WHO, 2010).

The captain of the team

Testosterone is the most significant male sex hormone; it is responsible for the development and maintenance of male sex organs and secondary sex characteristics, such as facial hair and vocal cord thickness and increased muscle mass and bone density.

Testosterone is also required for spermatogenesis. Testosterone is synthesized and secreted from the interstitial tissue of the testes (*Leydig's cells*), which are located in the testes adjacent to the seminiferous tubules. Leydig's cells secrete testosterone when stimulated by LH (Marieb & Hoehn, 2014, pp. 1093–1095; Smith & Walker, 2014).

Calling the plays

All embryos under 6 weeks' gestation are *ambisexual* (having neither male nor female sex characteristics). Between weeks 6 and 8 of gestation in a male fetus, the cortical tissue regresses and the medullary tissue enlarges, differentiating into the testes. Leydig's and Sertoli's cells then undergo rapid proliferation and testosterone secretion begins. It is not known if this action is autonomous or via stimulation from placental production of human chorionic gonadotropin (Considine, 2009). The presence of testosterone directly affects sexual differentiation in the fetus. With testosterone, fetal genitalia develop into a penis, scrotum, and testes; without testosterone, genitalia develop into a clitoris, vagina, and other female organs.

Androgenic activity normally causes the testes to descend into the scrotum via the inguinal canal by birth, though 1 in 20 full-term males babies are born with one or both of the testes undescended. This is called *cryptorchidism*. Most cases resolve by 6 months of age, but if it doesn't resolve naturally, the testis can be moved from the abdomen or inguinal canal into the scrotum with a procedure called orchidopexy (Hutson & Thorup, 2015).

Other key players

Male sex hormones are regulated by a system called the hypothalamic-pituitary-gonadal axis (HPG axis). The hypothalamus releases GnRH causing the anterior pituitary to release gonadotropins (LH and FSH). FSH stimulates spermatogenesis by indirectly causing higher concentrations of testosterone. LH binds to interstitial cells around the seminiferous tubules causing secretion of testosterone and the final steps of spermatogenesis. A high sperm count causes cells in the testis to produce a protein called inhibin. Rising levels of inhibin and testosterone create feedback to inhibit the release of GnRH and FSH. When spermatogenesis declines, the levels of inhibin fall (Smith & Walker, 2014).

No gonadotropins yet.

Time to grow from boy to man

During early childhood, gonadotropins aren't secreted and there is little circulating testosterone. Secretion of gonadotropins from the pituitary gland usually occurs between ages 11 and 14, and marks the onset of *puberty*. There is a dramatic rise in the production of testosterone (about 18-fold). This pubertal change in hormone

Male reproductive changes with aging

Physiologic changes in older men include reduced levels of circulating testosterone, which, in turn, may cause decreased libido, fatigue, depression, and erectile dysfunction. A reduced testosterone level also causes the testes to atrophy and soften and decreases sperm production and fertility (Malik et al., 2015).

Frequently, the prostate gland enlarges with age, and its secretions diminish. Seminal fluid also decreases in volume and becomes less viscous.

Sexual changes

During intercourse, older men may experience slower and weaker physiologic reactions. However, these changes don't necessarily lessen sexual satisfaction.

stimulates development and maturation of the primary sex characteristics—increase in the size of the penis and testes, start of spermatogenesis, libido, and full reproductive and erectile function—as well as secondary sex characteristics—hair growth, vocal cord changes, changes in lean muscle mass, increased bone density, and skin changes (thickening and change in sebaceous glands) (Peate & Nair, 2015).

Reaching the plateau

After a male achieves full physical maturity, usually by age 20, sexual and reproductive function remain fairly consistent throughout life. (See *Male reproductive changes with aging*.)

Female reproductive system

Unlike the male reproductive system, the female system is largely internal, housed within the pelvic cavity. It's composed of the vulva, vagina, cervix, uterus, fallopian tubes, and ovaries.

External genitalia

The external female genitalia is composed of the *mons pubis, labia majora, labia minora, clitoris, vaginal opening,* and adjacent structures and termed the vulva or pudendum (See *Female external genitalia.*) (Faiz et al., 2011; Peate & Nair, 2015).

Female external genitalia

The female external genitalia includes the vaginal opening, mons pubis, prepuce, clitoris, labia majora, labia minora, urethral meatus, and associated glands (Standring & Gray, 2008).

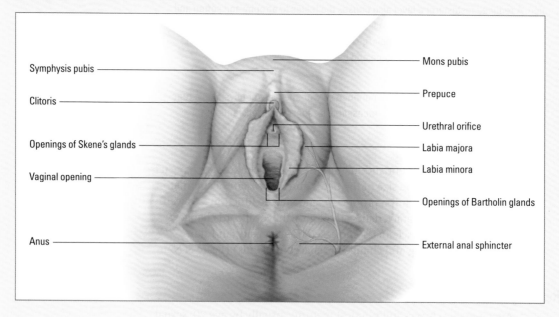

Symphysis pubis

Clitoris

Openings of Skene's glands

Vaginal opening

Anus

Mons pubis

Prepuce

Urethral orifice

Labia majora

Labia minora

Openings of Bartholin glands

External anal sphincter

At the top

The *mons pubis* is a rounded cushion of fatty and connective tissue, which, after puberty, is covered by skin and pubic hair in a triangular pattern over the symphysis pubis (the joint formed by the union of the pubic bones anteriorly).

Major league

The *labia majora* are two symmetrical raised folds of adipose and connective tissue and smooth muscle that border the vulva on either side, extending from the mons pubis to the perineum. The labia majora play a protective role for the vaginal and urethral orifices. After *menarche* (onset of menses), the outer surface of the labia is covered with pubic hair, while the inner aspect is pink and moist. The vulva contains a large concentration of sebaceous (oil) and apocrine (sweat) glands.

Minor...

The *labia minora* are two moist folds of connective mucosal tissue with a rich blood supply and are pink in appearance. Each labium minus lies longitudinally alongside the labium majus. Anteriorly they join to form the *prepuce*, a hoodlike covering over the clitoris. The posterior portion of the labia minora tapers to merge with the labia majora where they join to form the *fourchette*, a thin tissue fold along the anterior edge of the perineum (Marieb & Hoehn, 2014; Shier et al., 2015).

The *vestibule* is a recess enclosed by the labia minora and bounded anteriorly by the clitoris and posteriorly by the fourchette. Mucus-producing glands line the vestibule. The pair of lesser vestibular glands or paraurethral glands (*Skene's glands*) are located lateral to the urethral opening. The greater vestibular glands (*Bartholin's glands*) are located at the posteriolateral aspect of the inner vaginal orifice. The vestibule also contains the *urethral meatus*, the external opening through which urine leaves the female body, and, posterior to the urethra, the external opening of the vagina, the *vaginal introitus* (Marieb & Hoehn, 2014).

Small but sensitive

The *clitoris* is an approximately 2-cm organ that protrudes into the vulva anteriorly to the vestibule. The visible portion, just inferior to the prepuce, is called the glans clitoris. The body of the clitoris contains columns of erectile tissue (corpora cavernosa) and specialized sensory corpuscles; it is attached proximally by crura. The cavernosa and the bulbs of the vestibule become engorged with blood during the process of female sexual arousal (Shier et al., 2015).

Not too simple

A diamond-shaped region located between the pubic arch, coccyx, and ischial tuberosities, the *perineum* is a complex structure of muscles, blood vessels, fasciae, nerves, and lymphatics. The perineum contains the *central tendon*, the insertion site for the majority of the musculature that supports the pelvic floor (Marieb & Hoehn, 2014; Shier et al., 2015).

The labia are highly vascular and have many nerve endings—making them sensitive to pain, pressure, touch, sexual stimulation, and temperature extremes.

The internal genitalia

The female internal genitalia are specialized organs; their main functions are sexual and reproductive. (See *Structures of the female reproductive system*.)

Zoom in

Structures of the female reproductive system

The organs of the internal female reproductive system consist of the vagina, cervix, uterus, two fallopian tubes, and two ovaries (Standring & Gray, 2008).

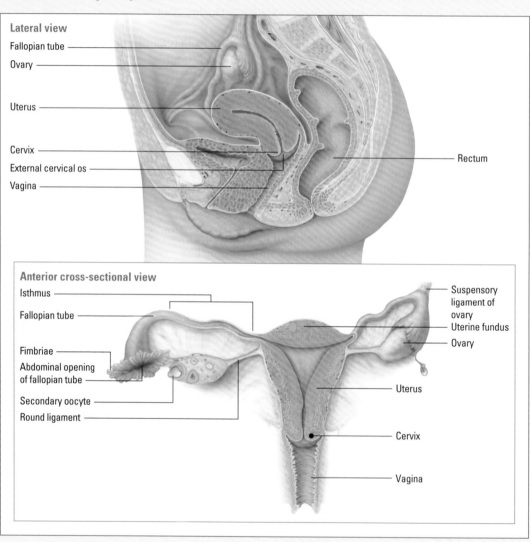

Lateral view
Fallopian tube
Ovary
Uterus
Cervix
External cervical os
Vagina
Rectum

Anterior cross-sectional view
Isthmus
Fallopian tube
Fimbriae
Abdominal opening of fallopian tube
Secondary oocyte
Round ligament
Suspensory ligament of ovary
Uterine fundus
Ovary
Uterus
Cervix
Vagina

Vagina

The *vagina* is an elastic fibromuscular tube approximately 9 cm long that lies at a 45-degree angle between the urethra and the rectum. The vaginal wall has three tissue layers: mucosal (inner stratified squamous epithelial lining), muscularis (two layers of smooth muscle), and adventitia (outer fibroelastic layer). The distal end opens into the vestibule, and the proximal end contains the *uterine cervix*. The uterine cervix connects the uterus to the vaginal vault. Four *fornices*, recesses in the vaginal wall, surround the cervix.

...three functions

The vagina has three main functions:
1. serving as the female organ of copulation
2. channeling blood discharged from the uterus during menstruation
3. serving as the birth canal during childbirth.

Blood and lymphatics

Blood is supplied to the vagina via the vaginal and uterine arteries, both of which are branches of the internal iliac artery.

Venous blood return is achieved via the vaginal venous plexus, through the uterine vein and finally drains into the internal iliac veins.

Lymphatic drainage of the vagina is via the iliac and superficial inguinal lymph nodes (Marieb & Hoehn, 2014; Shier et al., 2015).

Uterus

The *uterus* is a hollow, pear-shaped, muscular organ approximately 7 cm in length, 5 cm in length, and 2.5 cm in diameter; it is situated posteriosuperior to the bladder and anterior to the rectum. The typical uterus lies at nearly a 90-degree angle to the vagina. The uterus is constructed of three layers: a mucous membrane lining of columnar epithelium called the *endometrium*, a thick middle layer made of smooth muscle called the *myometrium*, and the outer serous membrane called the *perimetrium*. The upper two-thirds of the uterine body (uterine corpus) is called the *fundus*; the lower third is called the cervix. The uterus is supported by the broad ligament posteriorly and the round ligaments laterally.

Versatile organ

The endometrium and myometrium of the uterus respond to female sex hormones produced during the menstrual cycle and undergo a variety of changes during each component of the cycle to support a possible pregnancy. If pregnancy does not occur, hormone levels drop, and the nutrient-rich lining grown in preparation is sloughed off and menses occurs. If an egg is fertilized and an embryo implants in the endometrium, further hormones are secreted to sustain the elements necessary to support the pregnancy. In addition, during pregnancy, the elastic properties of the uterus allow it to expand to accommodate the growing fetus until term.

Cervix

The *uterine cervix* is the lower portion of the uterus that projects into the proximal vagina. The uterus is accessed from the vagina via a mucus-lined endocervical canal with a distal *external os* and proximal *internal os*. The cervix and vagina are supported by the uterosacral and cardinal ligaments (Marieb & Hoehn, 2014; Netter, Hansen, & Lambert, 2005; Shier et al., 2015; Waugh & Grant, 2014).

Permanent alterations

Childbirth permanently alters the cervix. In a female who hasn't delivered a child, the external os is usually a round opening about 3 mm in diameter; after the first vaginal childbirth, it becomes stellate or transverse with irregular edges. Also, prior to menopause, the uterine corpus is larger than the uterine cervix; after menopause, the size difference is reversed.

Fallopian tubes

Two *fallopian tubes*, approximately 10 to 13 cm in length and 0.7 cm in diameter, attach to the uterus laterally at the upper angles of the fundus. These narrow cylinders of fibromuscular tissue lined with columnar epithelium are the site of fertilization.

Riding the wave

The curved portion of the fallopian tube, called the *ampulla*, ends in the funnel-shaped *infundibulum*. Fingerlike

Fingerlike projections called *fimbriae* move in waves, sweeping the ovum from the ovary to the fallopian tube.

projections in the infundibulum, called *fimbriae*, move in waves that sweep the mature ovum from the ovary into the fallopian tube. Peristaltic contractions and ciliary movement help transport the egg through the tube.

Ovaries

The female gonads are called *ovaries*. They are paired organs located on either side of the uterus in the pelvic cavity. The ovaries are pale, solid, oval in shape, and approximately 3.5 cm long, 2 cm wide, and 1 cm thick during reproductive years; after menopause, the ovaries atrophy (shrink).

Limited supply

The ovaries' main functions are to produce sex hormones (estrogen, progesterone, and testosterone) and to produce and release eggs. At birth, each ovary contains approximately 1 to 2 million *follicles* (fluid-filled sacs that contain developing eggs called oocytes). By a process called atresia, many follicles degenerate, and by puberty, only 300,000 to 500,000 follicles remain. During the childbearing years, the body recruits several follicles each month; usually, only a single dominant oocyte matures and is extruded by the ovary each menstrual cycle; this is known as *ovulation*.

Hormones, menses, and oogenesis

Like the male body, the female body changes with age in response to hormonal control. When a female reaches the age of menstruation, the hypothalamus, ovaries, and pituitary gland (known as the HPO axis) secrete hormones—*estrogen, progesterone, FSH*, and *LH*—that trigger puberty and regulate the reproductive cycles and sexual responses of the female. (See *Events in the female reproductive cycle*, pages 288 to 290.)

Unlike the relatively short period for spermatogenesis males, the process of meiosis to create female sex cells (oogenesis) takes years. Females are born with the full supply of gametes (eggs) available to them in their ovaries. Oogonia (mitotic multiplication of diploid stem cells) and transformation into primary oocytes and primordial follicles occur in the fetal period, but primary oocytes do not progress past prophase I in the first meiotic division. The functionality of the female ovary is limited until hormonal stimulation at puberty; this means that all follicles recruited prior to puberty undergo *atresia* (cell death). Once puberty occurs, typically between 11 and 14 years of age, the secretion of FSH and LH allows some of these growing follicles to proceed to maturity and complete meiosis I. Meiosis II can only be completed if sperm penetrates the oocyte (Marieb & Hoehn, 2014; Shier et al., 2015; Waugh & Grant, 2014).

Mammary glands

Males and females have mammary glands—but they typically only function in the female.

The *mammary glands* are located in the subcutaneous tissue (breast tissue) of the thorax. They are specialized accessory organs of the female reproductive system designed to secrete milk following pregnancy. Although present in both sexes, they typically function only in the female.

Lobes, ducts, and drainage

Each mammary gland contains 15 to 25 lobes separated by fibrous connective tissue and fat. Within the lobes are lobules, clustered acini—tiny, saclike duct terminals that secrete milk during lactation. The alveolar glands empty into alveolar ducts that converge to form excretory (*lactiferous*) ducts and sinuses (*ampullae*), which store milk during lactation. These ducts drain onto the nipple surface through 15 to 20 openings. (See *The female breast.*) (Marieb & Hoehn, 2014; Netter et al., 2005; Shier et al., 2015).

(Text continues on page 291)

Zoom in

The female breast

The breasts are elevated tissue in the superficial fascia located on either side of the anterior chest wall over the pectoralis major and the anterior serratus muscles extending from sternum to axillae over the second to the sixth ribs. At the center of each breast lies the areola, a circular area of pigmented tissue that encompasses a single nipple. Smooth muscle tissue in the nipple responds to cold temperatures, friction, and sexual stimulation.

Support and separate

Each breast is composed of glandular, fibrous, and adipose tissue. Each mammary gland contains 15 to 25 lobes made up of lobules containing alveolar glands that secrete milk. The lobes are separated by dense connective and adipose tissues. Fibrous suspensory ligaments (*Cooper's ligaments*) support the breasts by attaching from the overlying dermis to the underlying pectoralis muscles.

Produce and drain

Acinus cells in the lobules of the mammary glands draw the ingredients needed to produce milk from the blood in surrounding capillaries. The milk is then secreted into the duct system of the breast and to the nipple as a result of contractions in surrounding myoepithelial cells. The myoepithelial cells are stimulated to contract by the female hormone oxytocin.

Sebaceous glands on the areolar surface, called *Montgomery's tubercles*, produce *sebum*, which lubricates the areolae and nipples during breast-feeding (Marieb & Hoehn, 2014; Netter et al., 2005; Shier et al., 2015).

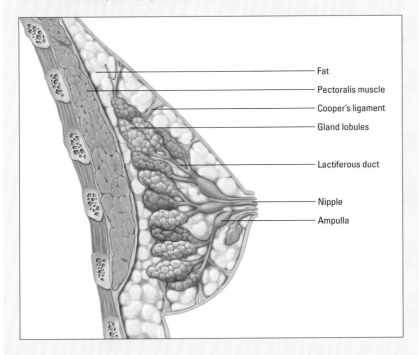

Fat

Pectoralis muscle

Cooper's ligament

Gland lobules

Lactiferous duct

Nipple

Ampulla

Now I get it!

Events in the female reproductive cycle

The three major types of change during the female reproductive cycle are ovarian, hormonal, and endometrial.

The female reproductive cycle involves a recurring concurrent series of complex changes in hormone production and secretion stimulating responses in the ovaries and endometrial lining of the uterus. The average female reproductive cycle usually occurs over 26 to 30 days. The first day of menstrual bleeding is considered the first day of the cycle. The cycle is generally described in three phases: menstrual, follicular (proliferative), and luteal (secretory).

Ovarian

• Ovarian changes begin on the first day of the menstrual cycle.
• As the cycle begins, low estrogen and progesterone levels in the bloodstream stimulate the hypothalamus to secrete gonadotropin-stimulating hormone (GnRH). In turn, GnRH stimulates the anterior pituitary gland to secrete follicle-stimulating hormone (FSH) and luteinizing hormone (LH).
• Follicle development (follicular phase) within the ovary is spurred by increasing levels of FSH and, to a lesser extent, LH.
• When the follicle matures, it produces a spike in estrogen, which stimulates a spike in the LH level, causing the follicle to rupture and release the ovum, thus initiating ovulation.
• After ovulation, the luteal phase begins. The collapsed follicle forms the corpus luteum, which (if fertilization doesn't occur) degenerates.

Hormonal

• The hypothalamus secretes GnRH triggering the anterior pituitary to secrete FSH and LH. FSH stimulates follicle growth leading to increased secretion of estrogen.
• Estrogen secretion peaks just before ovulation. This peak sets in motion the spike in LH levels, which triggers ovulation and stimulates development of the corpus luteum and secretion of progesterone.
• After ovulation, the corpus luteum is formed and begins to release progesterone, estrogen, and inhibin. The combination of estrogen and progesterone suppresses the hypothalamus and pituitary resulting in suppression of FSH secretion to prevent maturation of additional follicles in the same cycle.
• In the absence of pregnancy, both estrogen and progesterone levels decline as the corpus luteum degenerates, resulting in low levels of both hormones in the blood. This allows the hypothalamus and pituitary to resume activity, restarting the cycle.
• If fertilization does occur, the corpus luteum is supported by human chorionic gonadotropin secreted by the embryo.

Endometrial

• The endometrium is receptive to implantation of an embryo for only a short time in the reproductive cycle. Thus, it's no accident that the endometrium is most receptive about 7 days after the initiation of ovulation—just in time to receive a fertilized ovum.

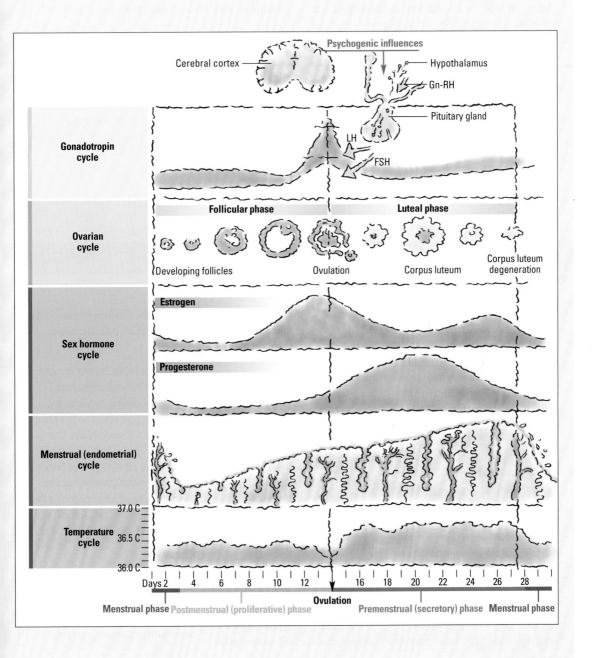

(continued)

Events in the female reproductive cycle *(continued)*

• In the first 5 days of the reproductive cycle (menstrual phase), the endometrium sheds its functional layer, leaving the basal layer (the deepest layer) intact. Menstrual flow consists of this detached layer and accompanying blood from degenerating capillaries.

• During the proliferative phase, the endometrium is stimulated by rising estrogen levels from the maturing follicle to begin regenerating the vascular functional layer rich in mucus-secreting glands.

• After ovulation, increased progesterone secretion stimulates conversion of the functional layer into an edematous secretory mucosa, which assists sperm motility through the cervix and uterus and is more receptive to implantation of the fertilized ovum.

• If implantation doesn't occur, LH levels fall; the corpus luteum degenerates; inhibin, estrogen, and progesterone levels drop; and the endometrium degenerates triggering menses and initiating a new cycle (Marieb & Hoehn, 2014; Netter et al., 2005; Shier et al., 2015; Waugh & Grant, 2014).

Senior moment

Female reproductive changes with aging

Women reach peak reproductive ability in their late 20s. By the late thirties, ovarian function is declining and fertility has started to decline. By age 50, there are very few follicles left in the ovaries. Declining estrogen and progesterone levels cause numerous physical changes in aging women. Significant emotional changes also take place during the transition from childbearing years to infertility Because women's breasts and internal and external reproductive structures are estrogen dependent, aging takes a more conspicuous toll on women than on men.

Ovaries
Ovulation becomes erratic in response to declining hormonal response in the 1 to 2 years prior to menopause. As the ovaries reach the end of their productive cycle, they become unresponsive to gonadotropic stimulation, and some cycles may be anovulatory, while some may result in ovulation of two or more oocytes. Once menopause is reached, the ovaries lose their ability to function as endocrine organs and atrophy.

Vulva
The glycogenated tissue of the vulva (and vagina) makes it extremely sensitive to the loss of estrogen. Decreased estrogen results in thinning and drying of the mucus membrane, defatting and flattening of the labia minora and majora, decreased moisture at the vestibule resulting in increased sensitivity, pruritus, and irritation. The tissues lose their elasticity, and the epidermis thins and the quantity of pubic hair decreases.

Vagina
As hormone levels decline, atrophic changes in the vaginal mucosa result in shortening, thinning, dry, less elastic, pale tissue. In this state, the vaginal mucosa is highly susceptible to abrasion and fissuring. In addition, the pH of vaginal secretions increases, making the vaginal environment more alkaline, changing the balance of natural flora and increasing the risk of vaginal infections.

Uterus
After menopause, the uterus shrinks rapidly to half its premenstrual weight. It continues to shrink until the organ reaches approximately one-fourth its premenstrual size.

Female reproductive changes with aging *(continued)*

The cervix atrophies and no longer produces mucus for lubrication; the endometrium and myometrium thin in response to absent estrogen stimulation and menses cease.

Pelvic support structures

Relaxation of the pelvic support commonly occurs in postreproductive women. Initial relaxation may occur after pregnancy and delivery, but clinical effects commonly go unnoticed until the process accelerates with menopausal estrogen depletion and loss of connective tissue elasticity and tone. Signs and symptoms may include pressure and pulling sensations in the area above the inguinal ligaments, lower backaches, a feeling of pelvic heaviness, and difficulty in rising from a sitting position. Urinary stress incontinence may also become a problem if urethrovesical ligaments weaken.

Breasts

In the breasts, glandular, supportive, and fatty tissues atrophy. As Cooper's ligaments lose their elasticity, the breasts become pendulous. The nipples decrease in size and become flat, and the inframammary ridges become more pronounced.

Big changes

During adolescence, the release of hormones causes a rapid increase in physical growth and spurs the development of secondary sex characteristics. These developmental changes usually occur between ages 11 and 14.

- Breast budding followed by progressive changes in size and structure until mature
- Hair growth at the pubis and axillae
- Changes in weight and body composition
- Increased in sebum and sweat production
- Menarche (the onset of menstruation)

Tanner staging is a scale that can be used to quantify the degree of sexual maturity an adolescent has achieved.

Supply exhausted

Cessation of menses usually occurs between ages 40 and 55. Depletion of oocytes prior to age 40 is called premature ovarian failure. While the pituitary gland still releases FSH and LH in menopausal women, the body has exhausted the supply of ovarian follicles that respond to these hormones, as a result, secretion of estrogen and progesterone are suppressed, endometrial tissue is no longer stimulated and menstruation ceases.

Menopause

A woman is considered to have reached menopause after menses are absent for 1 year. Prior menopause, a woman may experience several transitional years called perimenopause, during which many physiologic changes occur. (See *Female reproductive changes with aging*.)

The menstrual cycle may range from 22 to 34 days, although the typical cycle lasts 28 days.

The primary symptoms associated with menopause occur as a result of depleted estrogen, but progesterone depletion and testosterone depletion also affect women. The most common sequelae include vaginal dryness, vasomotor symptoms (hot flashes and night sweats), interrupted sleep, decreased libido, dry and thin skin, emotional lability, and loss of bone mass (Marieb & Hoehn, 2014; Shier et al., 2015).

Quick quiz

1. Spermatogenesis is the:
   A. growth and development of sperm into primary spermatocytes.
   B. division of spermatocytes into secondary spermatocytes.
   C. passage of sperm into the epididymis.
   D. entire process of sperm formation.

Answer: D. Spermatogenesis refers to the entire process of sperm formation, from the development of primary spermatocytes to the formation of fully functional spermatozoa.

2. The primary function of the scrotum is to:
   A. provide storage for newly developed sperm.
   B. maintain an appropriate temperature for the testes.
   C. deposit sperm in the female reproductive tract.
   D. secrete prostatic fluid.

Answer: B. The scrotum maintains an appropriate temperature for the testes, which is necessary for normal spermatozoa formation.

3. The main function of the ovaries is to:
   A. Regulate the buildup and shedding of the endometrium during the menstrual cycle.
   B. accommodate a growing fetus during pregnancy.
   C. produce ova and estrogen and progesterone.
   D. serve as the site of fertilization.

Answer: C. The main function of the ovaries is to produce ova, estrogen, and progesterone.

4. The corpus luteum forms and degenerates in which phase of the female reproductive cycle?
   A. Luteal
   B. Follicular
   C. Proliferative
   D. Secretory

Answer: A. The corpus luteum forms and degenerates in the luteal phase of the ovarian cycle.

5. Four hormones involved in the menstrual cycle are:
    A. LH, progesterone, estrogen, and testosterone.
    B. estrogen, FSH, LH, and androgens.
    C. estrogen, progesterone, LH, and FSH.
    D. gonadotropin-stimulating hormone, estrogen, progesterone, and testosterone.

Answer: C. The four hormones involved in the menstrual cycle are estrogen, progesterone, LH, and FSH.

Scoring

☆☆☆ If you answered all five questions correctly, congrats! You've hit a growth spurt in your knowledge.

☆☆ If you answered four questions correctly, good for you! You have a firm grasp of this complicated system.

☆ If you answered fewer than four questions correctly, keep trying! Like the reproductive system, your knowledge may just need some time to develop.

Selected References

Considine, R. V. (2009). Chapter 38: Fertilization, pregnancy and fetal differentiation. In R. A. Rhoades & D. R. Bell (Eds.), *Medical physiology: Principles for clinical medicine* (3rd ed.). Baltimore, MD: Lippincott Williams & Wilkins.

Cooper, T. G.; World Health Organization (WHO). (2010). WHO reference values for human semen characteristics. *Human Reproduction Update, 16*(5), 559.

Faiz, O., Blackburn, S., & Moffat, D. (2011). *Anatomy at a glance* (3rd ed.). Oxford, UK: Wiley-Blackwell.

Hutson, J. M., & Thorup, J. (2015). Evaluation and management of the infant with cryptorchidism. *Current Opinion in Pediatrics, 27*(4), 520–524. doi:10.1097/MOP.0000000000000237.

Malik, R. D., Lapin, B., Wang, C. E., Lakeman, J. C., & Helfand, B. T. (2015). Are we testing appropriately for low testosterone? Characterization of tested men and compliance with current guidelines. *Journal of Sexual Medicine, 12*(1), 66–75. doi:10.1111/jsm.12730.

Marieb, E. N., & Hoehn, K. N. (2014). *Human anatomy and physiology: Pearson new international edition* (9th ed.). Essex, UK: Pearson Education Limited.

Netter, F. H., Hansen, J. T., & Lambert, D. R. (2005). *Netter's clinical anatomy*. Carlstadt, NJ: Icon Learning Systems.

Peate, I., & Nair, M. (2015). *Anatomy and physiology for nurses at a glance*. Oxford, UK: Wiley-Blackwell.

Porter, R. S., Kaplan, J. L., & Merck & Co. (2011). *The Merck manual of diagnosis and therapy*. Whitehouse Station, NJ: Merck Sharp & Dohme Corp.

Shier, D., Butler, J., & Lewis, R. (2015). *Hole's essentials of human anatomy & physiology* (12th ed.). New York, NY: McGraw-Hill Education.

Smith, L. B., & Walker, W. H. (2014). The regulation of spermatogenesis by androgens. *Seminars in Cell and Developmental Biology, 30C*, 2–13. doi:10.1016/j.semcdb.2014.02.012.

Standring, S., & Gray, H. (2008). *Gray's anatomy: The anatomical basis of clinical practice.* Edinburgh, UK: Churchill Livingstone/Elsevier.

Waugh, A., & Grant, A. (2014). *Ross and Wilson: Anatomy and physiology in health and illness* (12th ed.). Edinburgh, UK: Churchill Livingstone/Elsevier.

Wilkins, R., Cross, S., Megson, I., Meredith, D. (Eds.). (2011). *Oxford handbook of medical sciences.* Oxford, UK: Oxford Medical Publications, Oxford University Press.

Chapter 17

Reproduction and lactation

Just the facts

In this chapter, you'll learn:

◆ the process of fertilization

◆ embryo and fetus development

◆ stages of labor

◆ the role of hormones in lactation.

Fertilization

Creation of a new human being begins with *fertilization*, the union of a *spermatozoon* and an *ovum* to form a single cell. After fertilization occurs, dramatic changes begin inside a woman's body and in the egg. The cells of the fertilized ovum begin dividing as the ovum travels to the *uterine cavity*, where it implants in the uterine lining. (See *How fertilization occurs*, page 296.)

One in a million

For fertilization to take place, however, a spermatozoon must first reach the ovum. Although a single ejaculation deposits several hundred-million spermatozoa, many are destroyed by acidic vaginal secretions. The only spermatozoa that survive are those that enter the *cervical canal*, where they're protected by *cervical mucus*.

I'd better get moving! For fertilization to take place, I have to reach the ovum.

Timing is everything

The ability of spermatozoa to penetrate the cervical mucus depends on the phase of the menstrual cycle at the time of transit.

Early in the cycle, estrogen and progesterone levels cause the mucus to thicken, making it more difficult for spermatozoa to pass through the cervix. During midcycle, however, when the mucus is relatively thin, spermatozoa can pass readily through the cervix. Later in the cycle, the cervical mucus thickens again, hindering spermatozoa passage.

How fertilization occurs

Fertilization begins when a spermatozoon is activated upon contact with the ovum. Here's what happens.

1. The spermatozoon, which has a covering called the *acrosome*, approaches the ovum.

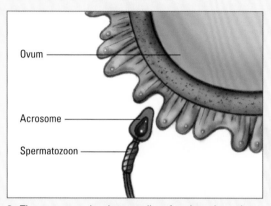

Ovum

Acrosome

Spermatozoon

2. The acrosome develops small perforations through which it releases enzymes necessary for the sperm to penetrate the protective layers of the ovum before fertilization.

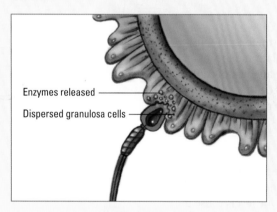

Enzymes released

Dispersed granulosa cells

3. The spermatozoon then penetrates the zona pellucida (the inner membrane of the ovum). This triggers the ovum's second meiotic division (following meiosis), making the zona pellucida impenetrable to other spermatozoa.

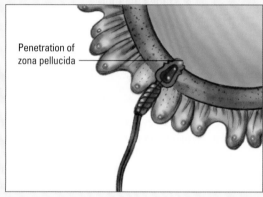

Penetration of zona pellucida

4. After the spermatozoon penetrates the ovum, its nucleus is released into the ovum, its tail degenerates, and its head enlarges and fuses with the ovum's nucleus. This fusion provides the fertilized ovum, called a *zygote*, with 46 chromosomes.

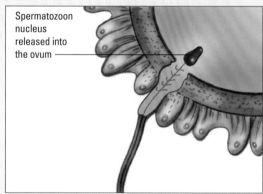

Spermatozoon nucleus released into the ovum

Help along the way

Spermatozoa travel through the female reproductive tract at a rate of several millimeters per hour by means of whiplike movements of the tail, known as *flagellar movements*.

After spermatozoa pass through the cervical mucus, however, the female reproductive system "assists" them on their journey with rhythmic contractions of the uterus that help them penetrate the fallopian tubes. Spermatozoa are typically viable (able to fertilize the ovum) for up to 2 days after ejaculation; however, they can survive in the reproductive tract for up to 4 days.

A zygote is "born"

Before a spermatozoon can penetrate the ovum, it must disperse the *granulosa* cells (outer protective cells) and penetrate the *zona pellucida*—the thick, transparent layer surrounding the incompletely developed ovum. Enzymes in the *acrosome* (head cap) of the spermatozoon permit this penetration. After penetration, the ovum completes its second meiotic division, and the zona pellucida prevents penetration by other spermatozoa.

The head of the spermatozoon then fuses with the ovum nucleus, creating a cell nucleus with 46 chromosomes. The fertilized ovum is called a *zygote*.

Pregnancy

Pregnancy starts with fertilization and ends with childbirth; on average, its duration is 38 to 40 weeks. During this period (called *gestation*), the zygote divides as it passes through the fallopian tube and attaches to the uterine lining via implantation. A complex sequence of *pre-embryonic*, *embryonic*, and *fetal* development transforms the zygote into a full-term fetus.

Making predictions

Because the uterus grows throughout pregnancy, uterine size serves as a rough estimate of gestation. The fertilization date is rarely known, so the woman's expected delivery date is typically calculated from the beginning of her last menses. The tool used for calculating delivery dates is known as *Nägele's rule.*

Here's how it works: If you know the first day of the last menstrual cycle, simply count back 3 months from that date and then add 7 days. For example, let's say that the first day of the last menses was April 29. Count back 3 months, which gets you to January 29, and then add 7 days for an approximate due date of February 5.

The tool used for calculating delivery dates is known as Nägele's rule.

Stages of fetal development

During pregnancy, the fetus undergoes three major stages of development:
1. pre-embryonic period
2. embryonic period
3. fetal period.

Rite of passage

The pre-embryonic phase starts with ovum fertilization and lasts for 2 weeks. As the zygote passes through the fallopian tube, it undergoes a series of *mitotic divisions,* or *cleavage.* (See *Pre-embryonic development.*)

Look, honey—it has your germ layers...

During the embryonic period (gestation weeks 3 through 8), the now-implanted structure starts to take on a human shape and is now called an *embryo.* Each germ layer—the *ectoderm, mesoderm,* and *endoderm*—eventually forms specific tissues in the embryo. (See *Embryonic development,* page 300.)

...my organ systems...

The organ systems form during the embryonic period. During this time, the embryo is particularly vulnerable to injury by maternal drug use, certain maternal infections, and other factors.

Memory jogger

To remember the three germ layers, keep in mind that *ecto* means outside, *meso* means middle, and *endo* means within.

Now I get it!

Pre-embryonic development

The pre-embryonic phase lasts from conception until approximately the end of the 2nd week of development.

Zygote formation...

As the fertilized ovum advances through the fallopian tube toward the uterus, it undergoes mitotic division, forming daughter cells, initially called *blastomeres,* that each contain the same number of chromosomes as the parent cell. The first cell division ends about 30 hours after fertilization; subsequent divisions occur rapidly.

The *zygote,* as it's now called, develops into a small mass of cells called a *morula,* which reaches the uterus at or around the 3rd day after fertilization. Fluid that amasses in the center of the morula forms a central cavity.

...into blastocyst

The structure is now called a *blastocyst.* The blastocyst consists of a thin trophoblast layer, which includes the blastocyst cavity, and the inner cell mass. The trophoblast develops into fetal membranes and the placenta. The inner cell mass later forms the embryo *(late blastocyst).*

Getting attached: Blastocyst and endometrium

During the next phase, the blastocyst stays within the zona pellucida, unattached to the uterus. The zona pellucida degenerates, and by the end of the first week after fertilization, the blastocyst attaches to the endometrium. The part of the blastocyst adjacent to the inner cell mass is the first part to become attached.

The trophoblast, in contact with the endometrial lining, proliferates and invades the underlying endometrium by separating and dissolving endometrial cells.

Letting it all sink in

During the next week, the invading blastocyst sinks below the endometrium's surface. The penetration site seals, restoring the continuity of the endometrial surface.

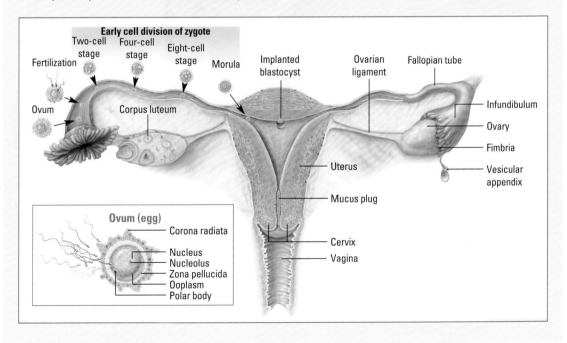

Now I get it!

Embryonic development

Each of the three germ layers—ectoderm, mesoderm, and endoderm—forms specific tissues and organs in the developing embryo.

Ectoderm

The ectoderm, the outermost layer, develops into the:
* epidermis
* nervous system
* pituitary gland
* tooth enamel
* salivary glands
* optic lens
* lining of the lower portion of the anal canal
* hair.

Mesoderm

The mesoderm, the middle layer, develops into:
* connective and supporting tissue
* the blood and vascular system
* musculature
* teeth (except enamel)
* the mesothelial lining of the pericardial, pleural, and peritoneal cavities
* the kidneys and ureters.

Endoderm

The endoderm, the innermost layer, becomes the epithelial lining of the:
* pharynx and trachea
* auditory canal
* alimentary canal
* liver
* pancreas
* bladder and urethra
* prostate.

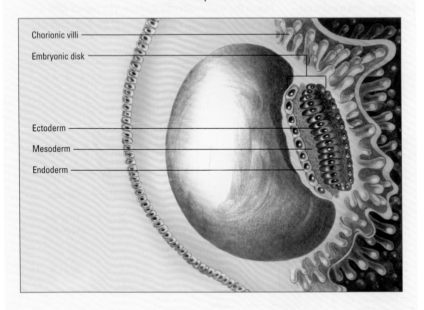

Chorionic villi

Embryonic disk

Ectoderm

Mesoderm

Endoderm

...and a very large head

During the *fetal* stage of development, which lasts from the 9th week until birth, the maturing fetus enlarges and grows heavier. (See *From embryo to fetus*.)

Now I get it!

From embryo to fetus

Significant growth and development take place within the first 3 months following conception, as the embryo develops into a fetus that nearly resembles a full-term newborn.

Month 1

At the end of the first month, the embryo has a definite form. The head, the trunk, and the tiny buds that will become the arms and legs are discernible. The cardiovascular system has begun to function, and the umbilical cord is visible in its most primitive form. The amniotic sac and placenta have begun to form. About 1/4in in length.

Month 2

During the 2nd month, the embryo—called a fetus from the 9th week—grows to 1" (2.5 cm) in length and weighs 1/30 oz (1 g). The head and facial features develop as the eyes, ears, nose, lips, tongue, and tooth buds form. The arms and legs also take shape. Although the gender of the fetus isn't yet discernible, all external genitalia are present. Cardiovascular function is complete, the umbilical cord has a definite form, and the neural tube is well formed. Around week 6, a heartbeat can usually be detected. At the end of the 2nd month, the fetus resembles a full-term newborn, except for size.

Month 3

During the 3rd month, the fetus grows to 4" (7.5 cm) in length and weighs 1 oz (28 g). Teeth and bones begin to appear, and the kidneys start to function. The fetus opens its mouth to swallow, grasps with its fully developed hands, and prepares for breathing by inhaling and exhaling amniotic fluid (although its lungs aren't functioning). At the end of the first *trimester* (the 3-month periods into which pregnancy is divided), the fetus's gender is distinguishable. The risk for miscarriage drops dramatically as the most critical development has taken place.

Months 4 to 9

Over the remaining 6 months, fetal growth continues as internal and external structures develop at a rapid rate. In the third trimester, the fetus stores the fats and minerals it will need to live outside the womb. At birth, the average full-term fetus measures 20" (51 cm) and weighs 7 to 71/2 lb (3 to 3.5 kg).

1 month

2 months

3 months

9 months

Two unusual features are noticeable during this stage:

1. The fetus's head is disproportionately large compared with its body. (This feature changes after birth as the infant grows.)
2. The fetus lacks subcutaneous fat. (Subcutaneous or white fat starts to accumulate shortly after birth.)

> The fetus isn't the only thing changing during pregnancy—the reproductive system undergoes changes as well.

Structural changes in the ovaries and uterus

Pregnancy changes the usual development of the *corpus luteum* (the ovum after ovulation, which secretes progesterone and small amounts of estrogen) and results in development of the decidua, amniotic sac and fluid, yolk sac, and placenta.

Corpus luteum

Normal functioning of the corpus luteum requires continual stimulation by *luteinizing hormone (LH)*. Progesterone produced by the corpus luteum suppresses LH release by the pituitary gland. If pregnancy occurs, the corpus luteum continues to produce progesterone until the placenta takes over. Otherwise, the corpus luteum atrophies 3 days before menstrual flow begins.

Hormone soup

With age, the corpus luteum grows less responsive to LH. For this reason, the mature corpus luteum degenerates unless stimulated by progressively increasing amounts of LH.

Pregnancy stimulates the placental tissue to secrete large amounts of human chorionic gonadotropin (HCG), which resembles LH and follicle-stimulating hormone (FSH), also produced by the pituitary gland. HCG prevents corpus luteum degeneration, stimulating the corpus luteum to produce large amounts of estrogen and progesterone needed to maintain the pregnancy during the first 3 months.

> HCG can be detected as early as 9 days after fertilization and can provide confirmation of pregnancy before the first menstrual period is missed.

HCG tells all

HCG can be detected as early as 9 days after fertilization and can provide confirmation of pregnancy even before the first menstrual period is missed.

The HCG level gradually increases, peaks at about 10 weeks' gestation, and then gradually declines.

Decidua

The *decidua* is the endometrial lining that undergoes the hormone-induced changes of pregnancy. Decidual cells secrete the following three substances:

- the hormone *prolactin*, which promotes lactation
- a peptide hormone, *relaxin*, which induces relaxation of the connective tissue of the symphysis pubis and pelvic ligaments and promotes cervical dilation
- a potent hormonelike fatty acid, *prostaglandin*, which mediates several physiologic functions.
  (See *Development of the decidua and fetal membranes*, page 304.)

Amniotic sac and fluid

The *amniotic sac*, enclosed within the chorion, gradually enlarges and surrounds the embryo. As it grows, the amniotic sac expands into the chorionic cavity, eventually filling the cavity and fusing with the chorion by 8 weeks' gestation.

This is the life—warm, buoyant, and protected!

A warm, protective sea

The amniotic sac and amniotic fluid serve the fetus in two important ways—one during gestation and the other during delivery. During gestation, the fluid provides the fetus with a buoyant, temperature-controlled environment that provides nutrition, and the first line of defense against pathogens. Later, it serves as a fluid wedge that helps open the cervix during birth.

Thanks, mom—I'll take it from here

Early in pregnancy, amniotic fluid comes chiefly from three sources:

1. fluid filtering into the amniotic sac from maternal blood as it passes through the uterus
2. fluid filtering into the sac from fetal blood passing through the placenta
3. fluid diffusing into the amniotic sac from the fetal skin and respiratory tract.
   Later in pregnancy, when the fetal kidneys begin to function, the fetus urinates into the amniotic fluid. Fetal urine then becomes the major component of amniotic fluid.

Daily gulps

Production of amniotic fluid from maternal and fetal sources compensates for amniotic fluid that's lost through the fetal GI tract. Normally, the fetus swallows up to several hundred milliliters of amniotic fluid each day. The fluid is absorbed into the fetal circulation from

Development of the decidua and fetal membranes

Specialized tissues support, protect, and nurture the embryo and fetus throughout its development. Among these tissues, the decidua and fetal membranes begin to develop shortly after conception.

Nesting place

During pregnancy, the endometrial lining is called the *decidua*. It provides a nesting place for the developing zygote and has some endocrine functions.

Network of blood vessels

The *chorion* is a membrane that forms the interior wall of the blastocyst. Vascular projections, called *chorionic villi*, arise from its periphery. As the *chorionic vesicle* enlarges, villi arising from the superficial portion of the chorion, called the *chorion laeve,* atrophy, leaving this surface smooth. Villi arising from the deeper part of the chorion, called the *chorion frondosum,* proliferate, projecting into the large blood vessels within the decidua basalis through which the maternal blood flows. The deepest part of the chorion will become the *chorionic plate of the placenta.*

Blood vessels form within the villi as they grow and connect with blood vessels that form in the chorion, in the body stalk, and within the body of the embryo. Blood begins to flow through this developing network of vessels as soon as the embryo's heart starts to beat.

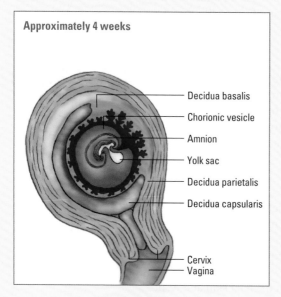

Approximately 4 weeks

Decidua basalis
Chorionic vesicle
Amnion
Yolk sac
Decidua parietalis
Decidua capsularis
Cervix
Vagina

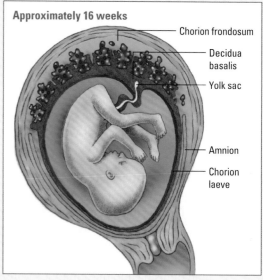

Approximately 16 weeks

Chorion frondosum
Decidua basalis
Yolk sac
Amnion
Chorion laeve

the fetal GI tract; some is transferred from the fetal circulation to the maternal circulation and excreted in maternal urine.

Yolk sac

Cells split off from the blastocyst and form a yolk sac. The *yolk sac* forms next to the endoderm; a portion of it is incorporated in the developing embryo and forms the GI tract. Another portion of the sac develops into primitive germ cells, which travel to the developing gonads and eventually form *oocytes* (precursors to ova) or *spermato-cytes* (precursors to spermatozoa), after gender is determined at fertilization.

No yolk—it's only temporary

During early embryonic development, the yolk sac also forms blood cells. Eventually, it atrophies and disintegrates.

Placenta

The flattened, disk-shaped *placenta,* using the umbilical cord as its conduit, provides nutrients to and removes wastes from the fetus from the 3rd month of pregnancy until birth. The placenta is formed from the chorion, its chorionic villi, and the adjacent decidua basalis.

The strongest link

The umbilical cord, which contains two arteries and one vein, links the fetus to the placenta. The umbilical arteries, which transport blood from the fetus to the placenta, take a spiral course on the cord, divide on the placental surface, and branch off to the chorionic villi. (See *Picturing the placenta.*)

Zoom in

Picturing the placenta

At term, the *placenta* (the spongy structure within the uterus from which the fetus derives nourishment) is flat, pancakelike, and round or oval. It measures approximately 22 cm in diameter and weighs about 1 lb. The maternal side is dark maroon and divided into lobules. The fetal side is shiny gray. After the birth, the placenta separates from the uterine wall and is expelled.

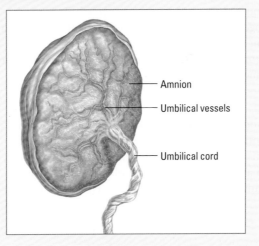

Amnion

Umbilical vessels

Umbilical cord

In a helpful vein

The placenta is a highly vascular organ. Large veins on its surface gather blood returning from the villi and join to form the single umbilical vein. The umbilical vein enters the cord, returning blood to the fetus.

Specialists on the job

The placenta contains two highly specialized circulatory systems:
- The *uteroplacental* circulation carries oxygenated arterial blood from the maternal circulation to the *intervillous spaces*—large spaces separating chorionic villi in the placenta. Blood enters the intervillous spaces from uterine arteries that penetrate the basal portion of the placenta; it leaves the intervillous spaces and flows back into the maternal circulation through veins in the basal portion of the placenta near the arteries.
- The *fetoplacental* circulation transports oxygen-depleted blood from the fetus to the chorionic villi through the umbilical arteries and returns oxygenated blood to the fetus through the umbilical vein.

Placenta takes charge

For the first 3 months of pregnancy, the corpus luteum is the main source of estrogen and progesterone—steroid hormones required during pregnancy. By the end of the 3rd month, however, the placenta produces most of the hormones; the corpus luteum persists but is no longer needed to maintain the pregnancy.

Hormones on the rise

Levels of estrogen and progesterone increase progressively throughout pregnancy. Estrogen stimulates uterine development to provide a suitable environment for the fetus. Progesterone, synthesized by the placenta from maternal cholesterol, reduces uterine muscle irritability and prevents spontaneous abortion of the fetus.

Keep those acids coming

The placenta also produces *human placental lactogen* (HPL), which resembles growth hormone. HPL stimulates maternal protein and fat metabolism to ensure a sufficient supply of amino acids and fatty acids for the mother and fetus. HPL also stimulates breast growth in preparation for lactation. Throughout pregnancy, HPL levels rise progressively.

Labor and the postpartum period

Childbirth is achieved through labor—the process by which uterine contractions expel the fetus from the uterus. When labor begins, these contractions become strong and regular. Eventually, voluntary bearing-down efforts supplement the contractions, resulting in delivery. When that occurs, presentation of the fetus takes one of a variety of forms. (See *Comparing fetal presentations*, page 308.)

Get ready to work! The contractions of labor are involuntary at first but are supplemented by voluntary efforts as birth approaches.

Onset of labor

The onset of labor results from several factors:

- The number of *oxytocin* (a pituitary hormone that stimulates uterine contractions) receptors on uterine muscle fibers increases progressively during pregnancy, peaking just before labor onset. This makes the uterus more sensitive to the effects of oxytocin.
- Stretching of the uterus over the course of the pregnancy initiates nerve impulses that stimulate oxytocin secretion from the posterior pituitary lobe.

Initiating start sequence

Near term, the fetal pituitary gland secretes more *adrenocorticotropic* hormone, which causes the fetal adrenal glands to secrete more *cortisol*. The cortisol diffuses into the maternal circulation through the placenta, heightens oxytocin and estrogen secretion, and reduces progesterone secretion. These changes intensify uterine muscle irritability and make the uterus even more sensitive to oxytocin stimulation.

Decline, diffuse, and contract

Declining progesterone levels convert *esterified arachidonic acid* into a nonesterified form. The nonesterified arachidonic acid undergoes biosynthesis to form prostaglandins, which, in turn, diffuse into the *uterine myometrium*, thereby inducing uterine contractions.

All systems go

As the cervix dilates, nerve impulses are transmitted to the central nervous system, causing an increase in oxytocin secretion from the pituitary gland. Acting as a positive feedback mechanism, increased oxytocin secretion stimulates more uterine contractions, which further dilate the cervix and lead the pituitary to secrete more oxytocin. Oxytocin secretions may also stimulate *prostaglandin* formation by the decidua.

Now I get it!

Comparing fetal presentations

Fetal presentations may be broadly classified as cephalic, breech, shoulder, or compound.

Cephalic	**Breech**	**Shoulder**	**Compound**

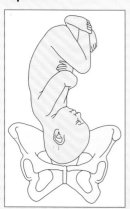 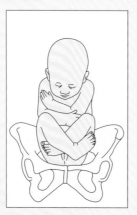 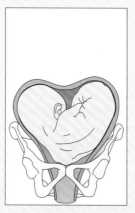 

In the *cephalic*, or head-down presentation, the position of the fetus may be classified further by the presenting skull landmark, such as vertex (the topmost part of the head), brow, sinciput (the forward, upper part of the skull), or face.

In the *breech,* or head-up, presentation, the position of the fetus may be further classified as frank (hips flexed, knees straight), complete (knees and hips flexed), footling (knees and hips of one or both legs extended), kneeling (knees flexed and hips extended), or incomplete (one or both hips extended and one or both feet or knees lying below the breech). Breech increases the risk of a loop in the umbilical cord that could cause injury if delivered vaginally.

Although a fetus may adopt one of several *shoulder* presentations, examination won't help a practitioner differentiate among them. Thus, all transverse positions are called *shoulder presentations*. This will likely require a cesarean delivery.

In a *compound* presentation, an extremity prolapses alongside the major presenting part so that two presenting parts appear at the pelvis at the same time.

Stages of labor

Childbirth can be divided into three stages. The duration of each stage varies according to the size of the uterus, the woman's age, and the number of previous pregnancies.

Stage 1—efface and dilate

The first stage of labor, in which the fetus begins its descent, is marked by cervical *effacement* (thinning) and *dilation*. Before labor begins, the cervix isn't dilated; by the end of the first stage, it has dilated fully. The first stage of labor can last from hours to days in primiparous women but is commonly significantly shorter for multiparous women. (See *Cervical effacement and dilation*, page 310.)

The first stage of labor can last from hours to days in primiparous women; the second stage lasts only about 45 minutes.

Stage 2—no turning back

The second stage of labor begins with full cervical dilation and ends with delivery of the fetus. During this stage, the amniotic sac ruptures as the uterine contractions increase in frequency and intensity. (The amniotic sac can also rupture before the onset of labor—during the first stage—and, although rare, sometimes ruptures after expulsion of the fetus.) As the flexed head of the fetus enters the pelvis, the mother's pelvic muscles force the head to rotate anteriorly and cause the back of the head to move under the symphysis pubis.

The curtain rises

As the uterus contracts, the flexed head of the fetus is forced deeper into the pelvis; resistance of the pelvic floor gradually forces the head to extend. As the head presses against the pelvic floor, vulvar tissues stretch and the anus dilates.

Both stages are typically much shorter for women who have delivered before.

The star appears

The head of the fetus now rotates back to its former position after passing through the vulvovaginal orifice. Usually, head rotation is lateral (external) as the anterior shoulder rotates forward to pass under the pubic arch. Delivery of the shoulders and the rest of the fetus follows. The second stage of labor can take minutes to hours. Typically longer for primiparous women and those who choose an epidural, it may be much shorter in multiparous women.

Curtain call

The third stage of labor starts immediately after childbirth and ends with placenta expulsion. After the neonate is delivered, the uterus continues to contract intermittently and grows smaller. The area of placental attachment also decreases. The placenta, which can't decrease in size, separates from the uterus, and blood seeps into the area of placental separation. The third stage of labor averages about 10 minutes in primiparous and multiparous women.

Cervical effacement and dilation

Cervical effacement and dilation are significant aspects of the first stage of labor.

Thinning walls

Cervical effacement is the progressive short-ening of the vaginal portion of the cervix and the thinning of its walls during labor as it's stretched by the fetus. Effacement is described as a percentage, ranging from 0% (noneffaced and thick) to 100% (fully effaced and paper-thin).

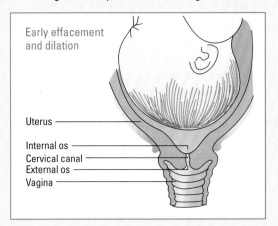

Early effacement and dilation

Uterus

Internal os
Cervical canal
External os
Vagina

Bigger exit

Cervical dilation refers to progressive enlarge-ment of the *cervical os* to allow the fetus to pass from the uterus into the vagina. Dilation ranges from less than 1 cm to about 10 cm (full dila-tion).

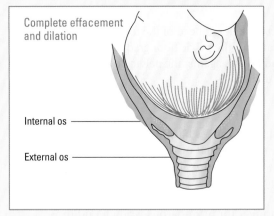

Complete effacement and dilation

Internal os

External os

Postpartum period

After childbirth, the reproductive tract takes about 6 weeks to heal and revert to its former condition during a process called *involution*. The uterus quickly grows smaller, with most of its involution taking place during the first 2 weeks after delivery.

It's all about the color

Postpartum vaginal discharge (lochia) persists for several weeks after childbirth:

- *Lochia rubra*, a bloody discharge, occurs immediately after delivery and lasts for 1 to 4 days postpartum.
- *Lochia serosa*, a pinkish brown, serous discharge, occurs from 5 to 7 days postpartum.
- *Lochia alba*, a grayish white or colorless discharge, appears from 1 to 3 weeks postpartum.

Lactation

Lactation (milk synthesis and secretion by the breasts) is governed by interactions involving four hormones:

- estrogen and progesterone, produced by the ovaries and placenta
- prolactin and oxytocin, produced by the pituitary gland under hypothalamic control.

Hormonal initiation of lactation

As gestation progresses, the production of estrogen and progesterone by the placenta increases, causing glandular and ductal tissue in the breasts to proliferate. After breast stimulation by estrogen and progesterone, prolactin causes milk secretion.

Priming the pump

Oxytocin from the posterior pituitary lobe causes contraction of specialized cells in the breast, producing a squeezing effect that forces milk down the ducts. Breast-feeding, in turn, stimulates prolactin secretion, resulting in a high prolactin level that induces changes in the menstrual cycle.

Opening the throttle

Progesterone and estrogen levels fall after delivery. With estrogen and progesterone no longer inhibiting prolactin's effects on milk production, the mammary glands start to secrete milk.

Pressing on the gas pedal

Nipple stimulation during breast-feeding results in transmission of sensory impulses from the nipples to the hypothalamus. If the nipples aren't stimulated by breast-feeding, prolactin secretion declines after delivery. Milk secretion continues as long as breast-feeding regularly stimulates the nipples. If breast-feeding stops, the stimulus for prolactin release is eliminated and milk production ceases.

Regular breast-feeding provides the stimulus necessary for continued milk production.

Breast-feeding and the menstrual cycle

During the postpartum period, the woman's high prolactin level inhibits FSH and LH release. If she doesn't breast-feed, her prolactin output soon drops, ending inhibition of FSH and LH production by the pituitary. Subsequently, cyclic release of FSH and LH occurs. (See *Breast milk composition.*)

A hold on ovulation

The menstrual cycle usually doesn't resume in breast-feeding women because prolactin inhibits the cyclic release of FSH and LH necessary for ovulation. This explains why a breast-feeding woman usually doesn't become pregnant.

The cycle comes full circle

Prolactin release in response to breast-feeding gradually declines, as does the inhibitory effect of prolactin on FSH and LH release. Consequently, ovulation and the menstrual cycle may resume. Pregnancy may occur after this, even if the woman continues to breast-feed.

Breast milk composition

The composition of breast milk undergoes various changes during the process of lactation.

First tastes

Initial feedings provide a thin, serous fluid called colostrum. Unlike mature breast milk, which has a bluish tinge, colostrum is yellow. Colostrum contains high concentrations of protein, fat-soluble vitamins, minerals, and immunoglobulins, which function as antibodies. Its laxative effect promotes early passage of meconium, the greenish black material that collects in the fetal intestines and forms the neonate's first stool. The breasts may contain colostrum for up to 96 hours after delivery.

From foremilk to hindmilk

Breast milk composition continues to change over the course of a feeding. The foremilk—the thin, watery milk secreted when a feeding begins—is low in calories but abounds in water-soluble vitamins. It accounts for about 60% of the total volume of a feeding. Next, hindmilk is released. The hindmilk, available several minutes after a breast-feeding session begins, has the highest concentration of calories; this helps to satisfy the neonate's hunger between feedings.

Quick quiz

1. Each of the three germ layers (ectoderm, mesoderm, and endoderm) forms specific tissues and organs in the developing:
    A. zygote.
    B. ovum.
    C. embryo.
    D. fetus.

Answer: C. Each of the three germ layers (ectoderm, mesoderm,
and endoderm) forms specific tissues and organs in the developing
embryo.

2. Which structure is responsible for protecting the fetus?
    A. Decidua
    B. Amniotic fluid
    C. Corpus luteum
    D. Yolk sac

Answer: B. Amniotic fluid, which provides a buoyant, temperature-
controlled environment, protects the fetus.

3. Progressive enlargement of the cervical os during labor is called:
    A. dilation.
    B. effacement.
    C. lactation.
    D. differentiation.

Answer: A. The progressive enlargement of the cervical os during
labor is called *dilation*.

4. The initial breast milk that's yellow in color is called:
    A. foremilk.
    B. hindmilk.
    C. colostrum.
    D. prolactin.

Answer: C. The initial breast milk that's yellow in color is called
colostrum.

Scoring

✰✰✰   If you answered all four questions correctly, fabulous! You're first-
rate with the physiology of fertilization.

✰✰   If you answered three questions correctly, excellent! You're looking
good in the areas of labor and lactation.

✰   If you answered fewer than three questions correctly, don't be
alarmed! Cycling through the information in the chapter again
should guarantee you won't reproduce those results.

Just for fun!

Unscramble the following words to discover the names of the three major stages of development during pregnancy. Then draw lines to link each stage with its particular characteristics.

REP-CRIBMONEY

— — —-

— — — — — — — —

BIMERCYNO

— — — — — — — —

ALEFT

— — — — —

A. Phase beginning at the 9th week of gestation and lasting until birth
B. Phase in which the zygote develops into a small mass of cells called a morula
C. Phase in which the zygote starts to taken on a human shape
D. Phase beginning with ovum fertilization and lasting for 2 weeks
E. Phase in which the fetus lacks subcutaneous fat
F. Phase in which the blastocyst attaches to the endometrium
G. Phase beginning at the 3rd week of gestation and lasting through the 8th week
H. Phase in which the zygote is called an embryo
I. Phase in which the fetus's head is disproportionately large compared with its body
J. Phase in which the organ systems form
K. Phase in which the zygote goes through a series of mitotic divisions

Answer: PRE-EMBRYONIC: B, D, F, K; EMBRYONIC: C, G, H, J; FETAL: A, E, I

Selected References

Alberts, B., Johnson, A., & Lewis J. (2002). Fertilization. *Molecular biology of the cell*. Retrieved from: http://www.ncbi.nlm.nih.gov/books/NBK26843/

Arey, L. (2016). Prenatal development. *Encyclopedia britannica*. Retrieved from https://www.britannica.com/science/prenatal-development

Fetal Positions for Birth. (2014). *Cleveland clinic*. Retrieved from http://my.clevelandclinic.org/health/diseases_conditions/hic_Am_I_Pregnant/hic_Labor_and_Delivery/hic_Fetal_Positions_for_Birth

Handwerger, S., & Freemark, M. (2000). The roles of placental growth hormone and placental lactogen in the regulation of human fetal growth and development. *Journal of Pediatric Endocrinology Metabolism, 13*(4). Retrieved from http://www.ncbi.nlm.nih.gov/pubmed/10776988

Kumar, P., & Magon, N. (2012). Hormones in pregnancy. *Nigerian Medical Journal: Journal of the Nigeria Medical Association, 53*(4), 179–183. http://doi.org/10.4103/0300-1652.107549

Labor, Delivery, and Postpartum Care. (2016). *Mayo clinic*. Retrieved from http://www.mayoclinic.org/healthy-lifestyle/labor-and-delivery/in-depth/stages-of-labor/art-20046545?pg=2

Stages Of Pregnancy & Fetal Development. (2016). *Cleveland clinic.* Retrieved from http://my.clevelandclinic.org/health/diseases_conditions/hic_Am_I_Pregnant/hic-fetal-development-stages-of-growth

Underwood, M., Gilbert, W., & Sherman, M. (2005). Amniotic fluid not just urine anymore. *Journal of Perinatology, 25,* 341–348. doi: 10.1038/sj.jp.7211290.

Wagner, C. Human milk and lactation. *Medscape.* Retrieved from http://emedicine.medscape.com/article/1835675-overview#a6

Yetter, J. (1998). Examination of the placenta. *American Family Physician, 57*(5). Retrieved from http://www.aafp.org/afp/1998/0301/p1045.html

Appendices

Practice makes perfect

1. The nurse is preparing a client for a cardiac workup. Which response by the nurse educates the client on the left atrium of the heart receiving most of its blood?
    A. "The left subclavian artery"
    B. "The internal carotid artery"
    C. "The left coronary artery"
    D. "The right coronary artery"

2. Which of the following actions indicate that the ventricles are receiving blood?
    A. Inspiration in the respiratory cycle
    B. Diastole of the cardiovascular cycle
    C. Expiration in the respiratory cycle
    D. Systole of the cardiovascular cycle

3. The nurse educates a cardiac client on which area of the heart that is most at risk for an infarction after prolonged occlusion of the right coronary artery?
    A. Anterior area of the heart
    B. Apical area of the heart
    C. Inferior area of the heart
    D. Lateral area of the heart

4. Which cardiac valve prevents the backflow of blood from the left ventricle into the left atrium?
    A. The aortic valve
    B. The mitral valve
    C. The pulmonic valve
    D. The tricuspid valve

5. The nurse educates a client on the effects of aerobic exercise in a health promotion class. Which symptom below does the nurse emphasis to the client may occur during exercise?
    A. A heart rate below the client's baseline
    B. A heart rate above the client's high range
    C. A decrease in the client's blood pressure
    D. A decrease in the client's respiratory effort

6. The nurse provides medication teaching at discharge. Signs and symptoms of high blood pressure should be reviewed for hormone?
    A. Angiotensin I
    B. Angiotensin II
    C. Renin
    D. Parathyroid hormone

7.   A decrease blood pressure is obtained in a client. The nurse understands this may be due to the hypothalamus secreting which of the following?
   A.   The hypothalamus secretes renin.
   B.   The hypothalamus secretes angiotensin.
   C.   The hypothalamus secretes epinephrine.
   D.   The hypothalamus secretes antidiuretic hormone (ADH).

8.   A nurse is analyzing a complete blood count for their client. Which of the following cells would be included on the report as an immature red blood cell (RBC)?
   A.   The CBC includes the B cell.
   B.   The CBC includes the T cell.
   C.   The CBC includes the reticulocyte.
   D.   The CBC includes the macrophage.

9.   A nurse knows the following are part of the structures included in the immune system in the human body?
   A.   The pancreas and liver
   B.   The adrenal glands and kidneys
   C.   The lymph nodes and thymus
   D.   The parotid glands and alveoli

10. The thymus gland provides which of the functions below to the human body?
   A.   The thymus gland is a reservoir for blood.
   B.   The thymus gland stores blood cells until they mature.
   C.   The thymus gland protects the body against ingested pathogens.
   D.   The thymus gland removes bacteria and toxins from the circulatory system.

11. A nurse working in the ICU knows that reinflating collapsed alveoli by ventilation improves oxygenation due to which of the following?
   A.   The alveoli require oxygen to remain viable.
   B.   The reinflated alveoli decrease the demand for oxygen.
   C.   The collapsed alveoli increase the demand for oxygen.
   D.   The gas exchange occurs in the alveolar membrane.

12. A nurse caring for a client with an injury to the hypothalamus is most likely at risk for which of the following symptoms?
  A. Experiencing uncontrollable seizures
  B. Obtaining an infection
  C. Complaining of fatigue and tiredness
  D. Struggling to have a normal sleep pattern

13. The nurse performs an assessment on a client who suffered a closed head injury and finds the client has memory loss and impaired hearing. Which brain lobe does the nurse suspect to have the most damage?
  A. The frontal lobe
  B. The occipital lobe
  C. The parietal lobe
  D. The temporal lobe

14. The nurse expects which connective tissues to enable their client's bones to move when the skeletal muscles contract?
  A. The tendons
  B. The ligaments
  C. The adipose tissues
  D. The nervous tissues

15. The nurse understands that osteoblast activity is necessary for which of the following functions in the human body?
  A. The bone formation function
  B. The estrogen production function
  C. The hematopoiesis function
  D. The muscle development function

16. Which of the following positively charged particles are closely packed in the nucleus of an atom?
  A. The electrons
  B. The neutrons
  C. The protons
  D. The subatomic particles

17. A nurse is reviewing triiodothyronine (T_3) and thyroxine (T_4) and knows these hormones affect which body processes?
  A. The metabolic rate
  B. The blood glucose level and glycogenesis
  C. The growth of bones, muscles, and other organs
  D. The bone resorption, calcium absorption, and blood calcium levels

18. The medulla of the adrenal gland releases which of the following hormones?
    A. The hormones epinephrine and norepinephrine
    B. The hormones insulin, glucagon, and somatostatin
    C. The hormones thyroxine (T_4), triiodothyronine (T_3), and calcitonin
    D. The hormones glucocorticoids, mineralocorticoids, and androgens

19. The nurse knows that the anterior pituitary gland secretes which hormones? Select all that apply.
    A. Prolactin
    B. Corticotropin
    C. Antidiuretic hormone (ADH)
    D. Follicle-stimulating hormone (FSH)
    E. Thyroid-stimulating hormone (TSH)

20. Which area of the human body is the thyroid gland located?
    A. The upper abdomen
    B. The inferior aspect of the brain
    C. The upper portion of the kidney
    D. The lower neck, anterior to the trachea

21. In the human body, the thyroid gland produces which of the following hormones? Select all that apply.
    A. The hormone amylase
    B. The hormone lipase
    C. The hormone calcitonin
    D. The hormone thyroxine (T_4)
    E. The hormone triiodothyronine (T_3)
    F. The hormone thyroid-stimulating hormone (TSH)

22. The skin acts as a protector of the inner body structures in the body. This is accomplished through which of the following actions?
    A. The repair surface wounds
    B. The impairing of the immune response
    C. The prevention of the secretion of sebum
    D. The halting of cell migration

23. A client complains that they frequently burn themselves while cooking because they cannot feel hot temperatures. Which lobe of the client's brain is most likely dysfunctional?
    A. The frontal lobe
    B. The occipital lobe
    C. The parietal lobe
    D. The temporal lobe

24. Emotional stress can cause sweating of the palms of the hands. Which of the following glands is responsible for this occurrence?

    A. The adrenal gland
    B. The eccrine gland
    C. The sebaceous gland
    D. The thyroid gland

25. During pregnancy, the fetal development undergoes different stages. Place the following stages of fetal development in order of their occurrence. Use all options.

A. The fetal period
B. The preembryonic period
C. The embryonic period

26. Which of the following functions are performed by the gallbladder in the human body?

    A. Recycling bile salts
    B. Storing and concentrating bile
    C. Removing bacteria from the blood
    D. Producing enzymes that assist with digestion

27. The heart pumps blood throughout the body. Which percentage of blood pumped is received by the kidneys each minute?

    A. 10%
    B. 20%
    C. 50%
    D. 70%

28. As blood circulates through the human body, it follows a path to become oxygenated and delivered throughout the body. Below, place the steps in order to trace the path of blood. Use all of the options.

| A. Right ventricle |
| B. Left ventricle |
| C. Pulmonary artery |
| D. Pulmonary vein |
| E. Left atrium |
| F. Aorta |
| |
| |
| |
| |
| |
| |

29. The nurse is educating a pregnant client in the clinic on how to estimate her due date. Using Nägele's rule, what date in October would the client be due if she states the first day of her last menstrual cycle was January 1?
    A. October 5th
    B. October 8th
    C. October 10th
    D. October 15th

30. When understanding the respiratory system, the nurse knows that excess bicarbonate retention may result in the client being in a state of:
    A. metabolic acidosis.
    B. respiratory acidosis.
    C. metabolic alkalosis.
    D. respiratory alkalosis.

31. When understanding nutrition in the human body, water-soluble vitamins are vital. Which of the following vitamins are water soluble?
    A. Vitamin A
    B. Vitamin C
    C. Vitamin E
    D. Vitamin K

32. In the human body, the prostate gland is a walnut-sized gland. This gland would be located in what area of the body?
   A. The area behind the bladder
   B. The area inside the scrotum
   C. The area along the posterior border of the testes
   D. The area under the bladder and around the urethra

33. The amniotic sac fills the chorionic cavity by:
   A. 4 weeks' gestation.
   B. 6 weeks' gestation.
   C. 8 weeks' gestation.
   D. 10 weeks' gestation.

34. When understanding the concept of fluid and electrolytes, the nurse knows that during osmosis, the fluid would move into which area below?
   A. Fluid would move from an area with more fluid to one with less fluid.
   B. Fluid would move from an area with less fluid to one with more fluid.
   C. Fluid would move from an area with less fluid to one with the same fluid.
   D. Fluid would move form an area with more fluid to one that also has more fluid.

35. When monitoring the intake and output for a client, the nurse knows that the average amount of urine that should be produced daily is?
   A. 200 to 700 mL
   B. 720 to 2,400 mL
   C. 2,500 to 4,500 mL
   D. 4,700 to 7,000 mL

36. A client is seen in the E.R. with possible respiratory distress. Based on the assessment findings documented, the nurse should suspect that the client has what alteration?

Progress notes	
07/1/16	Pt. wheezing. RR 44, BP 140/90, HR 104
1830	T 98.4°F. ABG results show pH 7.52
	Paco$_2$ 30 mm Hg, HCO$_3^-$ 26 mEq/L, and PO$_2$
	77 mm Hg
	— C.Smith, RN

   A. Metabolic acidosis
   B. Metabolic alkalosis
   C. Respiratory acidosis
   D. Respiratory alkalosis

37. In the human body, bone is created and destroyed. Which continuous process below is responsible for this action?
    A. The ossification process
    B. The formation process
    C. The remodeling process
    D. The classification process

38. Identify the carina in this illustration by placing an X in the proper location.

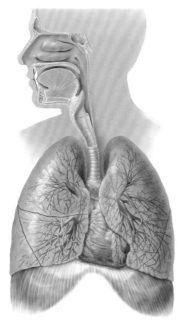

39. The connective tissue that covers and contours the spinal tissue and brain is known as:
    A. the pia mater.
    B. the dura mater.
    C. the endosteal dura.
    D. the meningeal dura.

40. The nurse is performing a neurological exam on a client and knows that the cranial nerve (CN) responsible for the sensation of taste is which of the following nerves?
    A. The facial nerve
    B. The trochlear nerve
    C. The olfactory nerve
    D. The hypoglossal nerve

41. Elderly patients have changes in nutrition due to physiologic changes. Which physiologic change below may affect this populations' nutrition? Select all that apply.
    A. Enhanced gag reflex
    B. Decreased biting force
    C. Decreased salivary flow
    D. Decreased renal function
    E. Diminished enzyme activity and gastric secretions

42. The transmission of sound vibrations occurs in the internal ear. Which structure below is responsible for this transmission?
    A. The auricle
    B. The vestibule
    C. The eustachian tube
    D. The tympanic membrane

43. Which of the following structure in the eye receives stimuli and then sends them to the brain?
    A. The lens
    B. The iris
    C. The sclera
    D. The retina

44. The nurse is analyzing a complete blood count. The nurse recognizes which type of blood cell indicates a possible infection?
    A. Platelets
    B. Thrombocytes
    C. Red blood cells (RBCs)
    D. White blood cells (WBCs)

45. The action of a body part moving backward and then forward is considered which of the following?
    A. Eversion and inversion
    B. Flexion and extension
    C. Pronation and supination
    D. Retraction and protraction

46. A client is experiencing problems with balance and fine and gross motor function. Identify the area of the client's brain that's malfunctioning by placing an X on that area of the illustration.

47. The acidity of the blood is determined by its acid-base pH. Which of the following pH values indicates acidity?
    A. 7.24
    B. 7.35
    C. 7.44
    D. 7.54

48. When providing education in a breastfeeding class, the nurse explains that two of the following hormones govern the lactation process. Which two hormones below are responsible for this process?
    A. Estrogen and progesterone
    B. Estrogen and corticotropin
    C. Growth hormone and progesterone
    D. Follicle-stimulating hormone (FSH) and estrogen

49. Autosomal recessive inheritance can be described as:
    A. the transmission of a dominant normal gene.
    B. the transmission of a recessive abnormal gene.
    C. the transmission of a dominant abnormal gene.
    D. the transmission of a recessive normal gene.

50. The movement of fluid or solutes from an area of higher concentration to an area of lower concentration is called an example of what type of action?
    A. The action of osmosis
    B. The action of diffusion
    C. The action of endocytosis
    D. The action of active transport

51. The nurse is assisting with the delivery of a fetus in breech position. Which graphic illustrates that position?

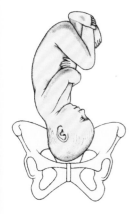 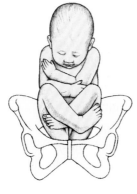  

Answers

1.   C. The left coronary artery, which splits into the anterior descending artery and the circumflex artery, is the primary source of blood for the left atrium. The left subclavian artery supplies blood to the arms, the internal carotid artery supplies blood to the head, and the right coronary artery supplies blood to the inferior wall of the heart.

2.   B. At the beginning of diastole, the semilunar valves close to prevent backflow of blood into the ventricles, and the mitral and tricuspid valves open, allowing blood to flow into the ventricles from the atria. Breathing patterns (inspiration and expiration) aren't related to blood flow.

3.   C. The right coronary artery supplies blood to the right atrium, part of the left atrium, most of the right ventricle, and the inferior portion of the left ventricle. Therefore, prolonged occlusion could produce an infarction in the inferior area. The right coronary artery doesn't supply blood to the anterior, lateral, or apical portions of the heart.

4.   B. The mitral valve prevents backflow of blood from the left ventricle into the left atrium. The aortic valve prevents backflow from the aorta into the left ventricle. The pulmonic valve prevents backflow from the pulmonary artery into the right ventricle. The tricuspid valve prevents backflow from the right ventricle into the right atrium.

5.   B. Stimulation of the sympathetic nervous system produces tachycardia and increased myocardial contractility. The other symptoms listed result from stimulation of the parasympathetic nervous system.

6.   B. Angiotensin II exerts a powerful constricting effect on arterioles, causing blood pressure to rise. Angiotensin I is converted to angiotensin II by angiotensin-converting enzyme. Renin is an enzyme that leads to the formation of angiotensin I. The main function of parathyroid hormone is to help regulate serum calcium levels.

7.   D. ADH acts on the renal tubules to promote water retention, which increases blood pressure. Although angiotensin, epinephrine, and renin also help increase blood pressure, they aren't stored in the hypothalamus.

8.   C. An immature RBC is called a *reticulocyte*. B cells, macrophages, and T cells are lymphocytes.

9.   C. The immune system includes the lymph nodes, thymus, spleen, and tonsils. The alveoli are part of the respiratory system. The adrenal glands are endocrine organs. The kidneys belong to the genitourinary system. The parotid glands, pancreas, and liver are part of the GI system.

10. B. Bone marrow produces immature blood cells (stem cells). Those that become lymphocytes migrate to the thymus for maturation (T lymphocytes). Lymphocytes are responsible for cell-mediated immunity. The spleen acts as a reservoir for blood cells. The tonsils shield against airborne and ingested pathogens. Lymph nodes remove bacteria and toxins from the bloodstream.

11. D. Gas exchange occurs in the alveolar membrane; therefore, collapsed alveoli decrease the surface area available for gas exchange, which decreases oxygenation of the blood. All alveoli, whether collapsed or not, receive oxygen and other nutrients from the bloodstream. Collapsed alveoli don't increase oxygen demand.

12. D. The hypothalamus helps regulate body temperature, appetite, water balance, pituitary secretions, emotions, and autonomic functions, including sleeping and waking cycles. Therefore, injury to the hypothalamus is most likely to cause difficulty sleeping. Neurotransmitter dysfunction is more likely to trigger seizures. A compromised immune system would make the patient tired and also place the patient at risk for infection.

13. D. The temporal lobe controls memory, hearing, and language comprehension. The frontal lobe influences thinking, planning, and judgment. The occipital lobe regulates vision. The parietal lobe interprets sensations.

14. A. Tendons are bands of connective tissue that attach muscles to bone, enabling them to move when skeletal muscles contract. Ligaments are fibrous connective tissue that bind bones to other bones. Adipose tissue is loose connective tissue that insulates the body. Nervous tissue isn't connective tissue.

15. A. Osteoblasts are bone-forming cells. Estrogen contributes to the development of bone tissue through calcium reuptake. Hematopoiesis (production of red blood cells) occurs in the bone marrow. Osteoblasts have no role in muscle development.

16. C. Protons are closely packed, positively charged particles within an atom's nucleus. Each element has a distinct number of protons. Electrons are negatively charged particles that orbit the nucleus in electron shells. Neutrons are uncharged, or neutral, particles in the

atom's nucleus. Protons, neutrons, and electrons are all subatomic particles, but each has a different charge.

17. A. T_3 and T_4 are thyroid hormones that affect metabolic rate. Bone resorption and increased calcium absorption are the principle effects of parathyroid hormone. Glucagon raises blood glucose levels and stimulates glycogenesis. The growth hormone somatotropin affects the growth of bones, muscles, and other organs.

18. A. The medulla of the adrenal gland releases epinephrine and norepinephrine. Glucocorticoids, mineralocorticoids, and androgens are released from the adrenal cortex. T_4, T_3, and calcitonin are secreted by the thyroid gland. The islet cells of the pancreas secrete insulin, glucagon, and somatostatin.

19. A, B, D, E. The anterior pituitary gland secretes corticotropin, FSH, TSH, and prolactin as well as growth hormone and luteinizing hormone. Inadequate secretion of corticotropin from the pituitary gland results in adrenal insufficiency. ADH is secreted by the posterior pituitary gland.

20. D. The thyroid gland resides in the lower neck, anterior to the trachea. The pancreas is in the upper abdomen. The pituitary gland is located in the inferior aspect of the brain. The adrenal glands are attached to the upper portion of the kidneys.

21. C, D, E. T_3, T_4, and calcitonin are all secreted by the thyroid gland. Amylase and lipase are enzymes produced by the pancreas. TSH is secreted by the pituitary gland.

22. A. The skin protects the body by intensifying normal cell replacement mechanisms to repair surface wounds. The epidermal layer of the skin contains Langerhans cells that enhance the immune response by helping lymphocytes process antigens entering the skin. Sebum is secreted by the sebaceous glands; the skin doesn't prevent its secretion. Migration and shedding of cells helps protect the skin.

23. C. The parietal lobe regulates sensory function, including the ability to sense hot or cold objects. The frontal lobe regulates thinking, planning, and judgment. The occipital lobe interprets visual stimuli. The temporal lobe regulates memory.

24. B. The eccrine glands, also known as *sweat glands*, secrete fluid on the palms and soles of the feet in response to emotional stress. The adrenal and thyroid glands produce hormones that control and affect body function. The sebaceous glands produce sebum, which helps protect the skin's surface; they're located in all areas of the skin except on the hands and soles of the feet.

25.

B.	The preembryonic period

C.	The embryonic period

A.	The fetal period

The preembryonic phase starts with ovum fertilization and lasts for 2 weeks. Weeks 3 through 8 encompass the embryonic period, during which time the developing zygote starts to take on a human shape and is called an *embryo*. The fetal stage of development lasts from week 9 until birth. During this period, the maturing fetus enlarges and grows heavier.

26. B. The gallbladder stores and concentrates bile produced by the liver. The pancreas produces enzymes that assist with digestion. The gallbladder isn't responsible for removing bacteria from blood. Bile salts are recycled by the liver.

27. B. The kidneys are highly vascular and receive about 20% of the blood pumped by the heart each minute.

28.

A.	Right ventricle

C.	Pulmonary artery

D.	Pulmonary vein

E.	Left atrium

B.	Left ventricle

F.	Aorta

Unoxygenated blood travels from the right ventricle through the pulmonic valve into the pulmonary arteries. After passing into the lungs, it travels to the alveoli, where it exchanges carbon dioxide for oxygen. The oxygenated blood returns via the pulmonary veins to the left atrium. It then passes through the mitral valve and into the left ventricle, where it is pumped out to the body via the aorta.

29. B. Nägele's rule calculates a due date by counting back 3 months from the first day of the last menstrual cycle and then adding 7 days. If the client's first day of her last period was January 1, she would be due on October 8.

30. C. Excess bicarbonate retention can result in metabolic alkalosis. Excess bicarbonate loss results in metabolic acidosis. Excess carbon dioxide retention results in respiratory acidosis. Excess carbon dioxide loss results in respiratory alkalosis.

31. B. Water-soluble vitamins include the B complex and C vitamins. Vitamins A, E, K, and D are fat-soluble vitamins.

32. D. The prostate gland lies under the bladder and surrounds the urethra. The vas deferens descends behind the bladder. The epididymis is located superior to and along the posterior border of the testes. The two sacs within the scrotum each contain a testis, an epididymis, and a spermatic cord.

33. C. The amniotic sac expands into the chorionic cavity, eventually filling the cavity and fusing with the chorion by 8 weeks' gestation.

34. A. In osmosis, fluid moves passively from an area with more fluid (and fewer solutes) to one with less fluid (and more solutes).

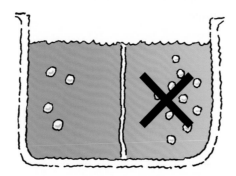

35. B. Total daily urine output averages 720 to 2,400 mL per day; however, this amount varies with fluid intake and climate.

36. D. Respiratory alkalosis results from hyperventilation. It's marked by a pH level above 7.45 and a concurrent decrease in partial pressure of arterial carbon dioxide ($Paco_2$) below 35 mm Hg. Metabolic alkalosis shows the same increase in pH but also an increase in bicarbonate level and a normal $Paco_2$. Acidosis of any type is characterized by a low pH level (below 7.35).

37. C. Remodeling is the continuous process whereby bone is created and destroyed. Formation refers to the development of bone. Ossification is bone hardening. Classification involves identifying bone according to its shape (long bone, short bone, or flat bone).

38. The carina is a ridge-shaped structure that's located at the level of the sixth or seventh thoracic vertebrae.

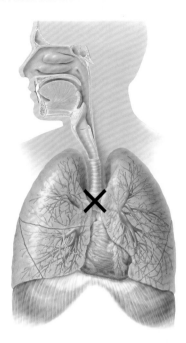

39. A. The pia mater is a continuous, delicate layer of connective tissue that covers and contours the spinal tissue and brain. The endosteal dura forms the periosteum of the skull and is continuous with the lining of the vertebral canal. The meningeal dura covers the brain, dipping between the brain tissues. The dura mater is leather-like tissue composed of the endosteal dura and meningeal dura.

40. A. The facial nerve (CN VII) controls taste and facial muscle movement. The olfactory nerve (CN I) is responsible for the sense of smell. The trochlear nerve (CN IV) controls extraocular eye movement. The hypoglossal nerve (CN XII) controls tongue movement.

41. B, C, D, E. Physiologic changes that affect nutrition in elderly patients include decreased renal function, decreased biting force, diminished enzyme activity and gastric secretions, and decreased salivary flow. Other changes include diminished intestinal activity, diminished sense of taste, and diminished gag reflex.

42. D. The tympanic membrane—which consists of layers of the skin, fibrous tissue, and a mucous membrane—transmits sound vibrations to the inner ear. The eustachian tube allows the pressure against inner and outer surfaces of the tympanic membrane to equalize, preventing rupture. The auricle is part of the external ear. The vestibule is the entrance to the inner ear.

43. D. The retina is the innermost coat of the eyeball; it receives visual stimuli and sends them to the brain. The lens refracts and focuses light onto the retina. The sclera helps maintain the size and form of the eyeball. The iris contains muscles and has an opening in the center for the pupil, which regulates light entry.

44. D. WBCs, which are classified as granulocytes or agranulocytes, participate in the body's defense and immune systems. RBCs transport oxygen and carbon dioxide to and from body tissues. Platelets, also known as *thrombocytes*, are involved in blood coagulation.

45. D. Retraction and protraction refer to backward and forward movement of a joint. Eversion and inversion involve moving a joint outward and inward. Flexion and extension involve increasing or decreasing a joint angle. Pronation and supination involve turning a body part downward or upward.

46. The cerebellum is the portion of the brain that controls balance and fine and gross motor function.

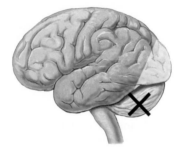

47. A. The body's pH control mechanism is so effective that blood pH stays within a narrow range: 7.35 to 7.45. Values below 7.35 indicate acidity; values above 7.45 indicate alkalinity.

48. A. The interaction of estrogen and progesterone with prolactin and oxytocin stimulates lactation. Corticotropin causes the fetus to secrete cortisol. FSH affects sexual development. Growth hormone influences growth and development.

49. B. Autosomal recessive inheritance refers to the transmission of a recessive abnormal gene—not transmission of a recessive normal gene. Autosomal dominant inheritance refers to the transmission of a dominant abnormal gene and not a dominant normal gene.

50. B. Diffusion is the movement of solutes from an area of higher concentration to one of lower concentration. Osmosis is the movement of fluid from an area of lower solute concentration into an area of higher solute concentration. Active transport involves using energy to move a substance across a cell membrane. Endocytosis is an active transport method in which a cell engulfs a substance.

51. B. In a breech presentation, the position of the head is up. In this breech, the buttocks are the presenting part and the hips and knees are flexed. In the cephalic presentation, the head is in the down position and enters the pelvis first. In the shoulder presentation, the fetus lies transverse and the shoulder is the presenting part. In the compound presentation, an extremity prolapses alongside the major presenting part so that two presenting parts appear at the pelvis at the same time.

abdomen: area of the body between the diaphragm and the pelvis

abduct: to move away from the midline of the body; the opposite of *adduct*

acetabulum: hip joint socket into which the head of the femur fits

acromion: bony projection of the scapula

adduct: to move toward the midline of the body; the opposite of *abduct*

adenoids: paired lymphoid structures located in the nasopharynx

adrenal gland: one of two secretory organs that lie atop the kidneys; consists of a medulla and a cortex

afferent neuron: nerve cell that conveys impulses from the periphery to the central nervous system; the opposite of *efferent neuron*

alveolus: small saclike dilation of the terminal bronchioles in the lung

ampulla: saclike dilation of a tube or duct

anterior: front or ventral; the opposite of *posterior* or *dorsal*

antibody: immunoglobulin produced by the body in response to exposure to a specific foreign substance (antigen)

antigen: foreign substance that causes antibody formation when introduced into the body

anus: distal end or outlet of the rectum

aorta: main trunk of the systemic arterial circulation, originating from the left ventricle and eventually branching into the two common iliac arteries

arachnoid: delicate middle membrane of the meninges

areola: pigmented ring around the nipple

arteriole: small branch of an artery

artery: vessel that carries blood away from the heart

arthrosis: joint or articulation

atrium: chamber or cavity

auricle: part of the ear that's attached to the head

axon: extension of a nerve cell that conveys impulses away from the cell body

bladder: membranous sac that holds secretions

bone: dense, hard connective tissue that composes the skeleton

bone marrow: soft tissue in the cancellous bone of the epiphyses; crucial for blood cell formation and maturation

bronchiole: small branch of the bronchus

bronchus: larger air passage of the lung

buccal: pertaining to the cheek

bursa: fluid-filled sac lined with synovial membrane

capillary: microscopic blood vessel that links arterioles with venules

carpal: pertaining to the wrist

cartilage: connective supporting tissue occurring mainly in the joints, thorax, larynx, trachea, nose, and ear

cecum: pouch located at the proximal end of the large intestine

celiac: pertaining to the abdomen

central nervous system: one of the two main divisions of the nervous system; consists of the brain and spinal cord

cerebellum: portion of the brain situated in the posterior cranial fossa, behind the brain stem; coordinates voluntary muscular activity

cerebrum: largest and uppermost section of the brain, divided into hemispheres

cilia: small, hairlike projections on the outer surfaces of some cells

cochlea: spiral tube that makes up a portion of the inner ear

colon: part of the large intestine that extends from the cecum to the rectum

condyle: rounded projection at the end of a bone

contralateral: on the opposite side; the opposite of *ipsilateral*

cornea: convex, transparent anterior portion of the eye

coronary: pertaining to the heart or its arteries

cortex: outer part of an internal organ; the opposite of *medulla*

costal: pertaining to the ribs

cricoid: ring-shaped cartilage found in the larynx

cutaneous: pertaining to the skin

deltoid: shaped like a triangle (as in the deltoid muscle)

dendrite: branching process extending from the neuronal cell body that directs impulses toward the cell body

dermis: skin layer beneath the epidermis

diaphragm: membrane that separates one part from another; the muscular partition separating the thorax and abdomen

diaphysis: shaft of a long bone

diarthrosis: freely movable joint

diencephalon: part of the brain located between the cerebral hemisphere and the midbrain

distal: far from the point of origin or attachment; the opposite of *proximal*

diverticulum: outpouching from a tubular organ such as the intestine

dorsal: pertaining to the back or posterior; the opposite of *ventral* or *anterior*

duct: passage or canal

duodenum: shortest and widest portion of the small intestine, extending from the pylorus to the jejunum

dura mater: outermost layer of the meninges

ear: organ of hearing

efferent neuron: nerve cell that conveys impulses from the central nervous system to the periphery; the opposite of *afferent neuron*

endocardium: interior lining of the heart

endocrine: pertaining to secretion into the blood or lymph rather than into a duct; the opposite of *exocrine*

epidermis: outermost layer of the skin; lacking vessels

epiglottis: cartilaginous structure overhanging the larynx that guards against entry of food into the lung

epiphyses: ends of a long bone

erythrocyte: red blood cell

esophagus: muscular canal that transports nutrients from the pharynx to the stomach

exocrine: pertaining to secretion into a duct; the opposite of endocrine

eye: one of two organs of vision

fallopian tube: one of two ducts extending from the uterus to the ovary

fontanel: incompletely ossified area of a neonate's skull

foramen: small opening

fossa: hollow or cavity

fundus: base of a hollow organ; the part farthest from the organ's outlet

gallbladder: excretory sac lodged in the visceral surface of the liver's right lobe

ganglion: cluster of nerve cell bodies found outside the central nervous system

genitalia: reproductive organs; may be external or internal

gland: organ or structure body that secretes or excretes substances

glomerulus: compact cluster; the capillaries of the kidney

gonad: sex gland in which reproductive cells form

heart: muscular, cone-shaped organ that pumps blood throughout the body

hemoglobin: protein found in red blood cells that contains iron and transports oxygen

hormone: substance secreted by an endocrine gland that triggers or regulates the activity of an organ or cell group

hyoid: shaped like the letter U; the U-shaped bone at the base of the tongue

hypothalamus: structure in the diencephalon that secretes vasopressin and oxytocin

ileum: distal part of the small intestine extending from the jejunum to the cecum

incus: one of three bones in the middle ear

inferior: lower; the opposite of *superior*

intestine: portion of the GI tract that extends from the stomach to the anus

intima: innermost structure

ipsilateral: on the same side; the opposite of *contralateral*

jejunum: one of three portions of the small intestine; connects proximally with the duodenum and distally with the ileum

joint: fibrous, cartilaginous, or synovial connection between bones

kidney: one of two urinary organs on the dorsal part of the abdomen

labia: Latin-derived term meaning "lip"; usually used to describe external female genitalia; part of the vulva

lacrimal: pertaining to tears

larynx: voice organ; joins the pharynx and trachea

lateral: pertaining to the side; the opposite of *medial*

leukocyte: white blood cell

ligament: band of white fibrous tissue that connects bones

liver: large gland in the right upper abdomen; divided into four lobes

lobe: defined portion of any organ, such as the liver or brain

lobule: small lobe

lumbar: pertaining to the area of the back between the thorax and the pelvis

lungs: organs of respiration found in the chest's lateral cavities

lymph: watery fluid in lymphatic vessels

lymph node: small oval structure that filters lymph, fights infection, and aids hematopoiesis

lymphocyte: white blood cell; the body's immunologically competent cells

malleolus: projections at the distal ends of the tibia and fibula

malleus: tiny hammer-shaped bone in the middle ear

mammary: pertaining to the breast

manubrium: upper part of the sternum

meatus: opening or passageway

medial: pertaining to the middle; the opposite of *lateral*

mediastinum: middle portion of the thorax between the pleural sacs that contain the lungs

medulla: inner portion of an organ; the opposite of *cortex*

membrane: thin layer or sheet

metacarpals: bones of the hand located between the wrist and the fingers

metatarsals: bones of the foot located between the tarsal bones and the toes

muscle: fibrous structure whose contraction initiates movement

myocardium: thick, contractile layer of muscle cells that forms the heart wall

nares: nostrils

nephron: structural and functional unit of the kidney

nerve: cordlike structure consisting of fibers that convey impulses from the central nervous system to the body

neuron: nerve cell

neutrophil: white blood cell that removes and destroys bacteria, cellular debris, and solid particles

occiput: back of the head

olfactory: pertaining to the sense of smell

ophthalmic: pertaining to the eye

ossicle: small bone, especially of the ear

ovary: one of two female reproductive organs found on each side of the lower abdomen, next to the uterus

palate: roof of the mouth

pancreas: secretory gland in the epigastric and hypogastric regions

parotid: located near the ear (as in the parotid gland)

patella: floating bone that forms the kneecap

pectoral: pertaining to the chest or breast

pelvis: funnel-shaped structure; lower part of the trunk

pericardium: fibroserous sac that surrounds the heart and the origin of the great vessels

phalanx: one of the tapering bones that makes up the fingers and toes

pharynx: tubular passageway that extends from the base of the skull to the esophagus

phrenic: pertaining to the diaphragm

pia mater: innermost covering of the brain and spinal cord

pituitary gland: gland attached to the hypothalamus that stores and secretes hormones

plantar: pertaining to the sole of the foot

plasma: colorless, watery fluid portion of lymph and blood

platelet: small, disk-shaped blood cell necessary for coagulation

pleura: thin serous membrane that encloses the lung

plexus: network of nerves, lymphatic vessels, or veins

pons: portion of the brain that lies between the medulla and the mesencephalon

popliteal: pertaining to the back of the knee

posterior: back or dorsal; the opposite of *anterior* or *ventral*

pronate: to turn the palm downward; the opposite of *supinate*

prostate: male gland that surrounds the bladder neck and urethra

proximal: situated nearest the center of the body; the opposite of *distal*

pupil: circular opening in the iris of the eye through which light passes

reflex: involuntary action

renal: pertaining to the kidney

scrotum: skin pouch that houses the testes and parts of the spermatic cords

semen: male reproductive fluid

sphenoid: wedged-shaped bone at the base of the skull

spleen: highly vascular organ between the stomach and the diaphragm

stapes: tiny stirrup-shaped bone in the middle ear

sternum: long, flat bone that forms the middle portion of the thorax

stomach: major digestive organ, located in the right upper abdomen

striated: marked with parallel lines such as striated (skeletal) muscle

superior: higher; the opposite of *inferior*

supinate: to turn the palm of the hand upward; the opposite of *pronate*

symphysis: growing together; a type of cartilaginous joint in which fibrocartilage firmly connects opposing surfaces

synapse: point of contact between adjacent neurons

systole: contraction of the heart muscle

talus: anklebone

tarsus: instep

tendon: band of fibrous connective tissue that attaches a muscle to a bone

testis: one of two male gonads that produce semen

thyroid: secretory gland located at the front of the neck

tibia: shinbone

tongue: chief organ of taste, found in the floor of the mouth

trachea: nearly cylindrical tube in the neck, extending from the larynx to the bronchi, that serves as a passageway for air

turbinate: shaped like a cone or spiral; a bone located in the posterior nasopharynx

ureter: one of two thick-walled tubes that transport urine to the bladder

urethra: small tubular structure that drains urine from the bladder

uterus: hollow, internal female reproductive organ in which the fertilized ovum is implanted and the fetus develops

uvula: tissue projection that hangs from the soft palate

vagina: sheath; the canal in the female extending from the vulva to the cervix

valve: structure that permits fluid to flow in only one direction

vein: vessel that carries blood to the heart

vena cava: one of two large veins that returns blood from the peripheral circulation to the right atrium

ventral: pertaining to the front or anterior; the opposite of *dorsal* or *posterior*

ventricle: small cavity, such as one of several in the brain or one of the two lower chambers of the heart

venule: small vessel that connects a vein and capillary plexuses

vertebra: any of the 33 bones that make up the spinal column

viscera: internal organs

xiphoid: sword-shaped; the lower portion of the sternum

Locating body cavities

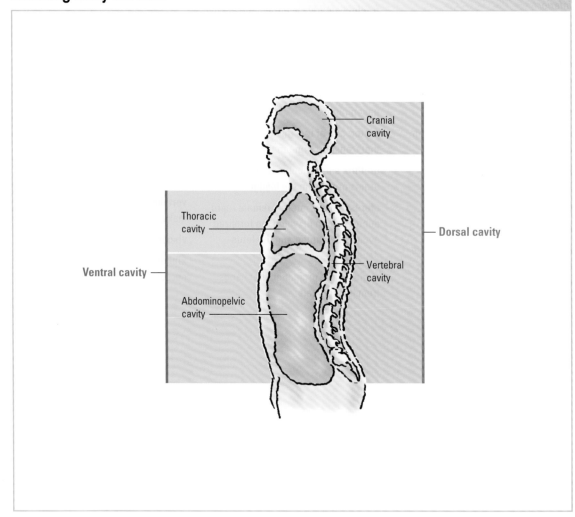

Cranial cavity

Thoracic cavity

Dorsal cavity

Ventral cavity

Vertebral cavity

Abdominopelvic cavity

Inside the cell

Cytoplasm (protoplasm that surrounds the nucleus) ——

Cell membrane (encloses the cell) ——

Mitochondrion (production site of adenosine triphosphate—cellular energy) ——

Centriole (takes part in cell division) ——

Endoplasmic reticulum (transports protein and lipid components) ——

Ribosomes (sites for protein synthesis) ——

Golgi complex (processes and packages protein) ——

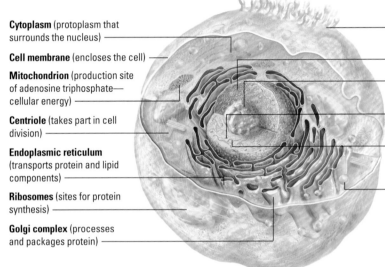

Microvilli (increase surface size of the cell)

Nucleus (brain of the cell)

Nucleolus (site of ribosomal RNA synthesis)

Ribonucleic acid (transfers genetic information to ribosomes)

Chromatin (complex of DNA, RNA, and protein that makes up chromosomes)

Lysosome (contains digestive enzymes)

A close look at the skin

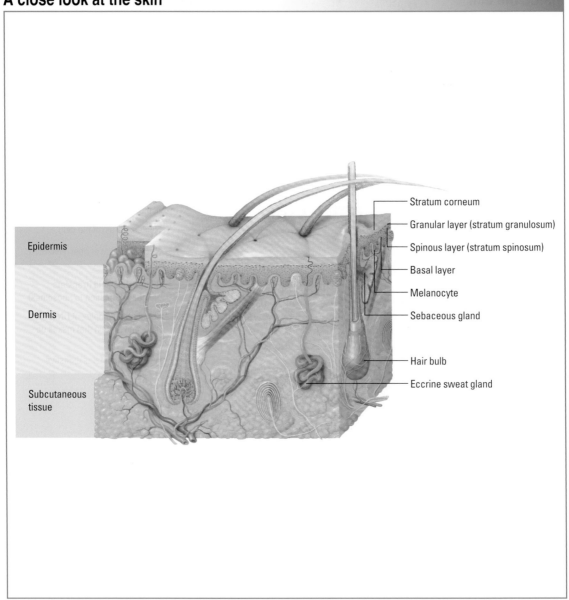

Epidermis

Dermis

Subcutaneous tissue

Stratum corneum

Granular layer (stratum granulosum)

Spinous layer (stratum spinosum)

Basal layer

Melanocyte

Sebaceous gland

Hair bulb

Eccrine sweat gland

A close look at the nail

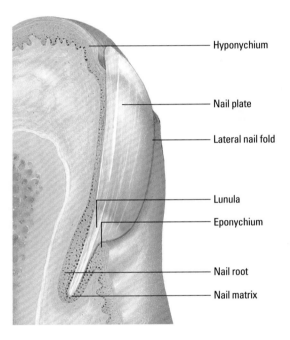

- Hyponychium
- Nail plate
- Lateral nail fold
- Lunula
- Eponychium
- Nail root
- Nail matrix

Viewing the major skeletal muscles

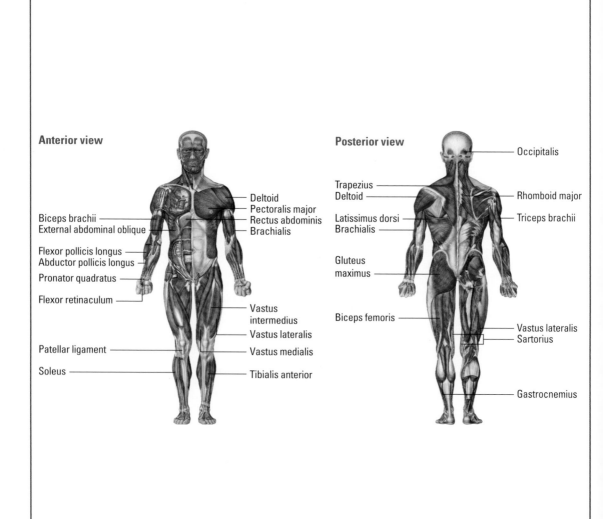

Anterior view

Biceps brachii
External abdominal oblique

Flexor pollicis longus
Abductor pollicis longus

Pronator quadratus

Flexor retinaculum

Patellar ligament

Soleus

Deltoid
Pectoralis major
Rectus abdominis
Brachialis

Vastus
intermedius
Vastus lateralis

Vastus medialis

Tibialis anterior

Posterior view

Trapezius
Deltoid

Latissimus dorsi
Brachialis

Gluteus
maximus

Biceps femoris

Occipitalis

Rhomboid major

Triceps brachii

Vastus lateralis
Sartorius

Gastrocnemius

Muscle structure up close

Periosteum
A tough, fibrous connective tissue that covers the surface of bones, rich in sensory nerves, responsible for healing fractures

Epimysium
Fibrous tissue enveloping the entire muscle and continuous with the tendon

Fascicle
A group of fibers that have been bound by perimysium

Endomysium
A delicate connective tissue that surrounds each muscle fiber

Sarcomere
Portion of muscle fibers found between two Z-lines

Tendon
A dense, fibrous connective tissue that is continuous with the periosteum and attaches muscle to the bone

Belly
Thick contractile portion (or body) of the muscle

Perimysium
Fibrous tissue that extends inward from epimysium, surrounding bundles of muscle. Each bundle bound by perimysium is called a fascicle.

Z-line
Boundary of two sarcomeres

H-zone
Consists of stacks of myosin filaments

Muscle fibers
Long, cylindrical, multinucleated cells with striations

Myosin and actin
Thick and thin filaments active during muscle contraction

Viewing the major bones

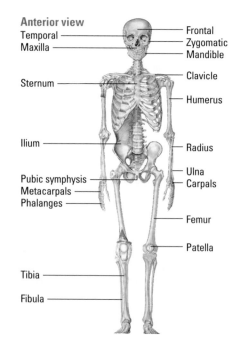

Anterior view

Temporal
Maxilla

Sternum

Ilium

Pubic symphysis
Metacarpals
Phalanges

Tibia

Fibula

Frontal
Zygomatic
Mandible

Clavicle

Humerus

Radius

Ulna
Carpals

Femur

Patella

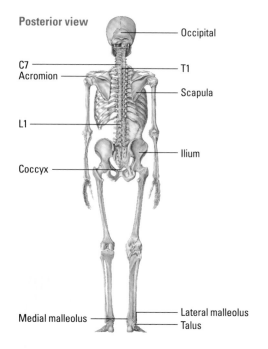

Posterior view

C7
Acromion

L1

Coccyx

Medial malleolus

Occipital

T1

Scapula

Ilium

Lateral malleolus
Talus

Parts of a neuron

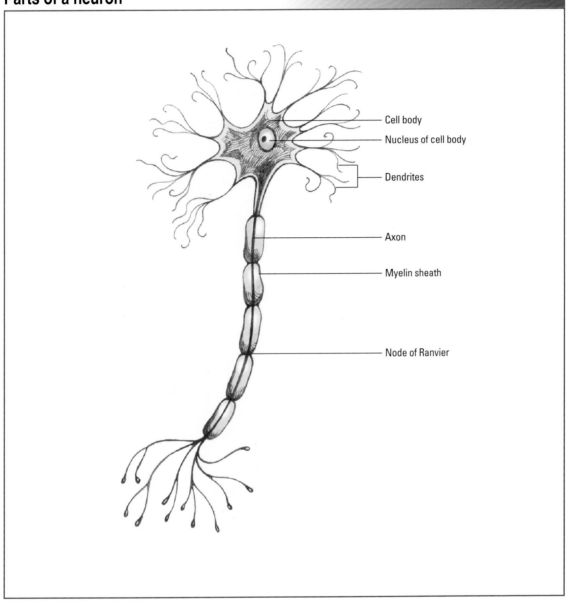

Cell body

Nucleus of cell body

Dendrites

Axon

Myelin sheath

Node of Ranvier

The reflex arc

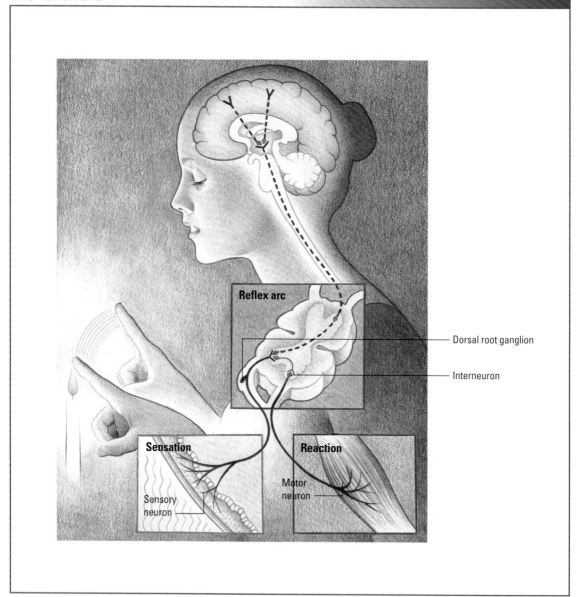

Reflex arc

Dorsal root ganglion

Interneuron

Sensation

Sensory neuron

Reaction

Motor neuron

A close look at major brain structures

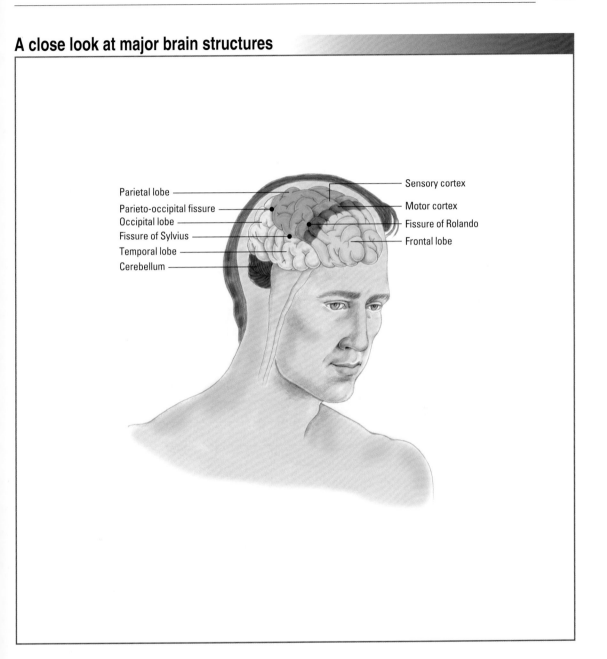

Parietal lobe

Parieto-occipital fissure

Occipital lobe

Fissure of Sylvius

Temporal lobe

Cerebellum

Sensory cortex

Motor cortex

Fissure of Rolando

Frontal lobe

The limbic system

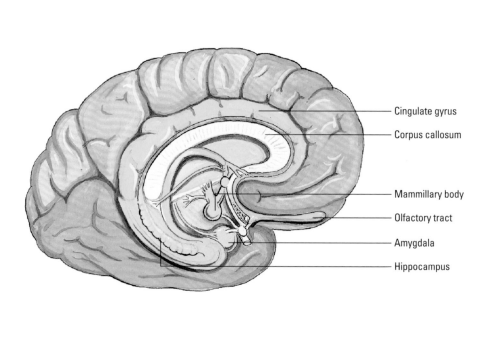

Cingulate gyrus

Corpus callosum

Mammillary body

Olfactory tract

Amygdala

Hippocampus

Arteries of the brain

Inferior view

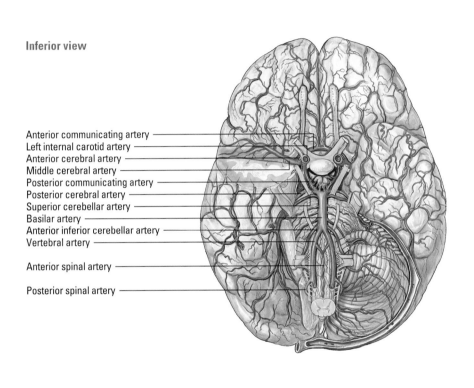

Anterior communicating artery
Left internal carotid artery
Anterior cerebral artery
Middle cerebral artery
Posterior communicating artery
Posterior cerebral artery
Superior cerebellar artery
Basilar artery
Anterior inferior cerebellar artery
Vertebral artery

Anterior spinal artery

Posterior spinal artery

A look inside the spinal cord

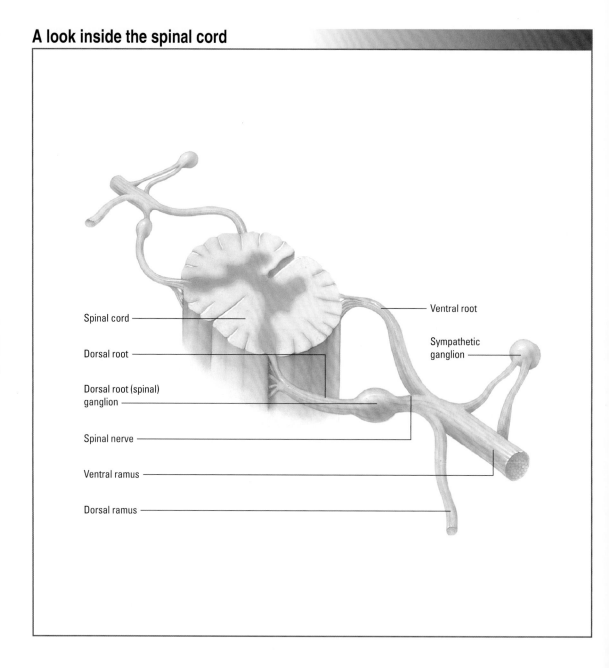

Spinal cord

Dorsal root

Dorsal root (spinal)
ganglion

Spinal nerve

Ventral ramus

Dorsal ramus

Ventral root

Sympathetic
ganglion

Major neural pathways

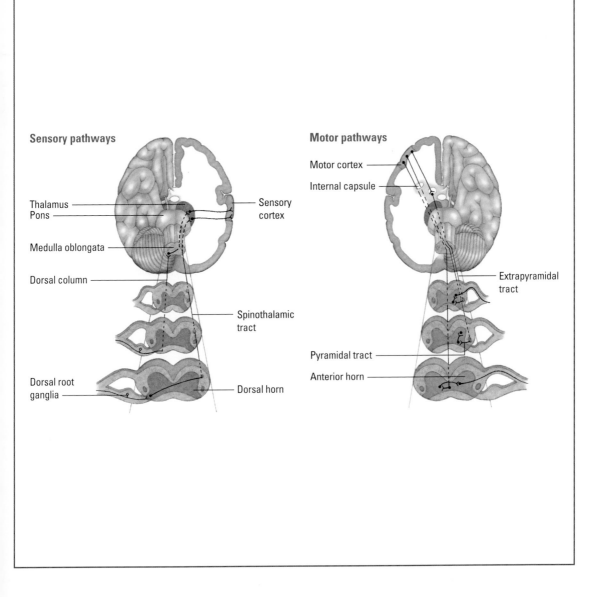

Sensory pathways

Thalamus
Pons
Medulla oblongata
Dorsal column
Dorsal root ganglia

Sensory cortex
Spinothalamic tract
Dorsal horn

Motor pathways

Motor cortex
Internal capsule
Pyramidal tract
Anterior horn

Extrapyramidal tract

The spinal nerves

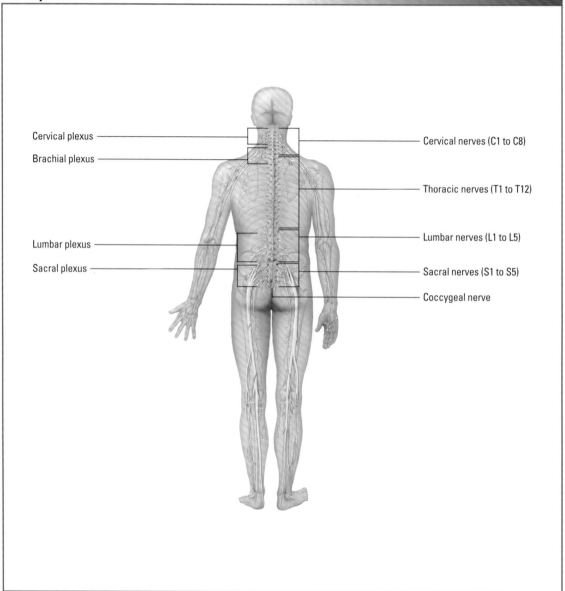

Cervical plexus

Brachial plexus

Lumbar plexus

Sacral plexus

Cervical nerves (C1 to C8)

Thoracic nerves (T1 to T12)

Lumbar nerves (L1 to L5)

Sacral nerves (S1 to S5)

Coccygeal nerve

Exit points for the cranial nerves

Inferior view

Oculomotor (CN III). *Motor*: extraocular eye movement (superior, medial, and inferior lateral), pupillary constriction, upper eyelid elevation

Trochlear (CN IV). *Motor*: extraocular eye movement (superior oblique muscles of the eyes)

Abducens (CN VI). *Motor*: extraocular eye movement (lateral)

Acoustic (CN VIII). *Sensory*: hearing, sense of balance

Glossopharyngeal (CN IX). *Motor*: swallowing movements; *Sensory*: sensations of throat, taste receptors (posterior one-third of tongue)

Facial (CN VII). *Sensory:* taste receptors (anterior two-thirds of tongue); *Motor*: facial muscle movement, including muscles of expression (those in the forehead and around the eyes and mouth)

Optic (CN II). *Sensory*: vision

Olfactory (CN I). *Sensory*: smell

Trigeminal (CN V). *Sensory*: transmitting stimuli from face and head, corneal reflex; *Motor*: chewing, biting, and lateral jaw movements

Hypoglossal (CN XII). *Motor*: tongue movement

Vagus (CN X). *Motor*: movement of palate, swallowing, gag reflex, activity of the thoracic and abdominal viscera, such as heart rate and peristalsis; *Sensory:* sensations throat, larynx, and thoracic and abdominal viscera (heart, lungs, bronchi, and GI tract)

Spinal accessory (CN XI). *Motor*: shoulder movement, head rotation

A close look at tears

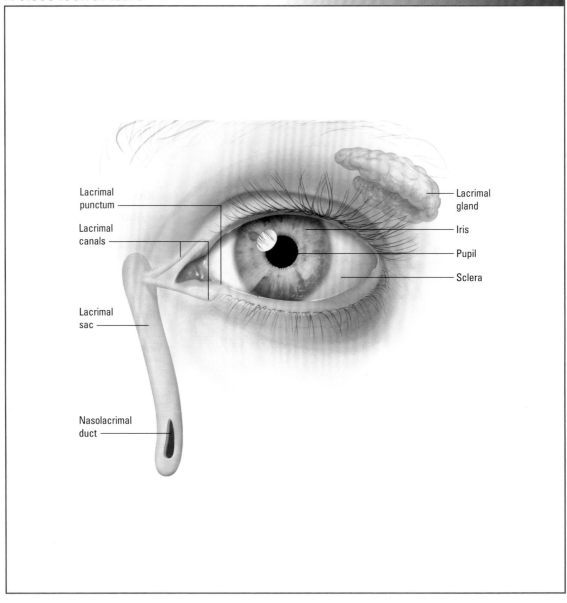

Lacrimal punctum

Lacrimal canals

Lacrimal sac

Nasolacrimal duct

Lacrimal gland

Iris

Pupil

Sclera

Intraocular structures

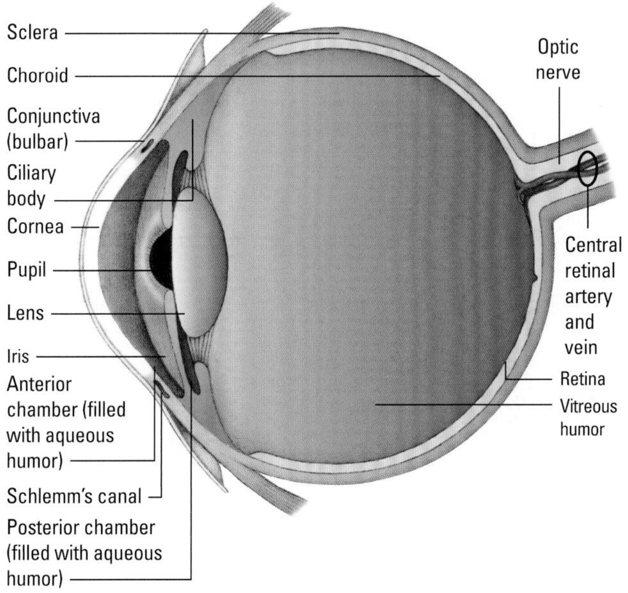

Sclera

Choroid

Conjunctiva (bulbar)

Ciliary body

Cornea

Pupil

Lens

Iris

Anterior chamber (filled with aqueous humor)

Schlemm's canal

Posterior chamber (filled with aqueous humor)

Optic nerve

Central retinal artery and vein

Retina

Vitreous humor

Retinal structures: A closer view

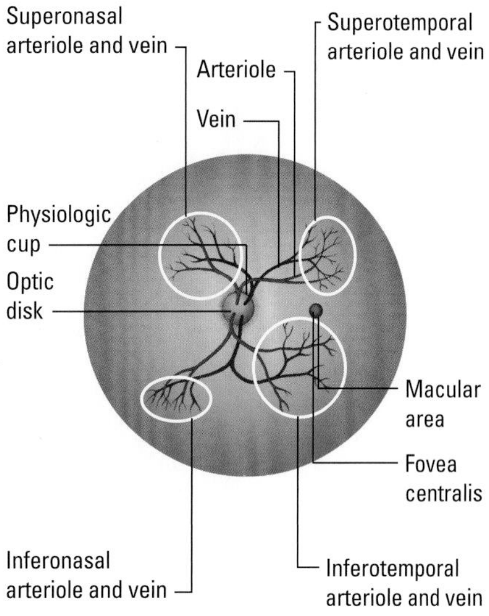

Superonasal arteriole and vein

Arteriole

Vein

Physiologic cup

Optic disk

Superotemporal arteriole and vein

Macular area

Fovea centralis

Inferonasal arteriole and vein

Inferotemporal arteriole and vein

Ear structures

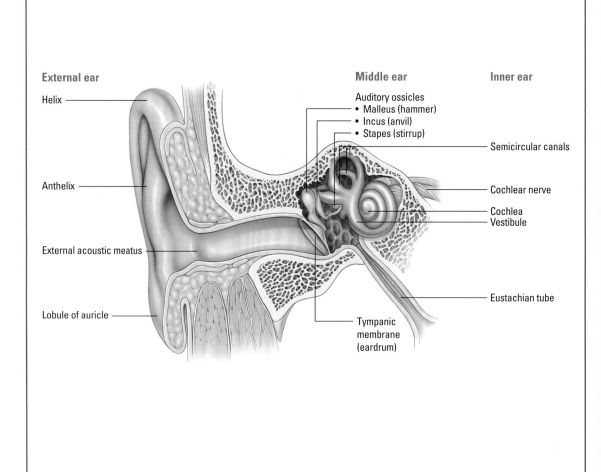

External ear

Helix

Anthelix

External acoustic meatus

Lobule of auricle

Middle ear

Auditory ossicles
- Malleus (hammer)
- Incus (anvil)
- Stapes (stirrup)

Tympanic membrane (eardrum)

Inner ear

Semicircular canals

Cochlear nerve

Cochlea
Vestibule

Eustachian tube

Components of the endocrine system

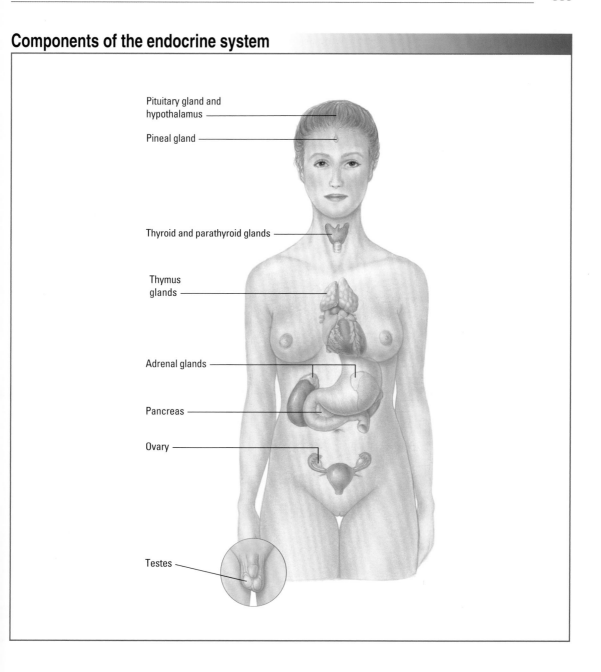

Pituitary gland and hypothalamus

Pineal gland

Thyroid and parathyroid glands

Thymus glands

Adrenal glands

Pancreas

Ovary

Testes

Inside the heart

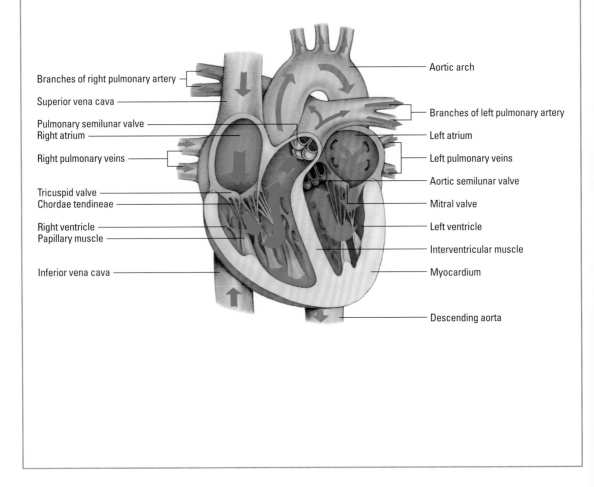

Branches of right pulmonary artery

Superior vena cava

Pulmonary semilunar valve
Right atrium

Right pulmonary veins

Tricuspid valve
Chordae tendineae

Right ventricle
Papillary muscle

Inferior vena cava

Aortic arch

Branches of left pulmonary artery

Left atrium

Left pulmonary veins

Aortic semilunar valve

Mitral valve

Left ventricle

Interventricular muscle

Myocardium

Descending aorta

Cardiac conduction system

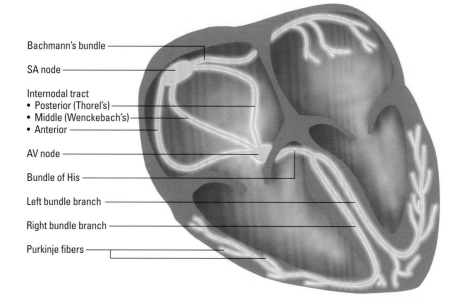

Bachmann's bundle

SA node

Internodal tract
• Posterior (Thorel's)
• Middle (Wenckebach's)
• Anterior

AV node

Bundle of His

Left bundle branch

Right bundle branch

Purkinje fibers

Major blood vessels

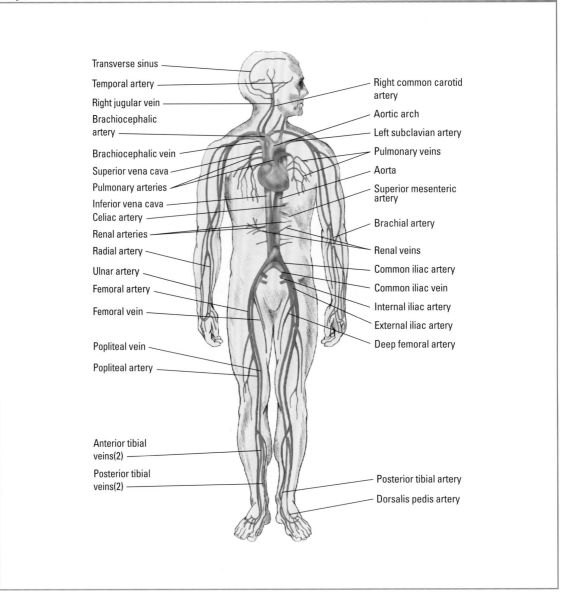

Transverse sinus

Temporal artery

Right jugular vein

Brachiocephalic artery

Brachiocephalic vein

Superior vena cava

Pulmonary arteries

Inferior vena cava

Celiac artery

Renal arteries

Radial artery

Ulnar artery

Femoral artery

Femoral vein

Popliteal vein

Popliteal artery

Anterior tibial veins(2)

Posterior tibial veins(2)

Right common carotid artery

Aortic arch

Left subclavian artery

Pulmonary veins

Aorta

Superior mesenteric artery

Brachial artery

Renal veins

Common iliac artery

Common iliac vein

Internal iliac artery

External iliac artery

Deep femoral artery

Posterior tibial artery

Dorsalis pedis artery

Vessels that supply the heart

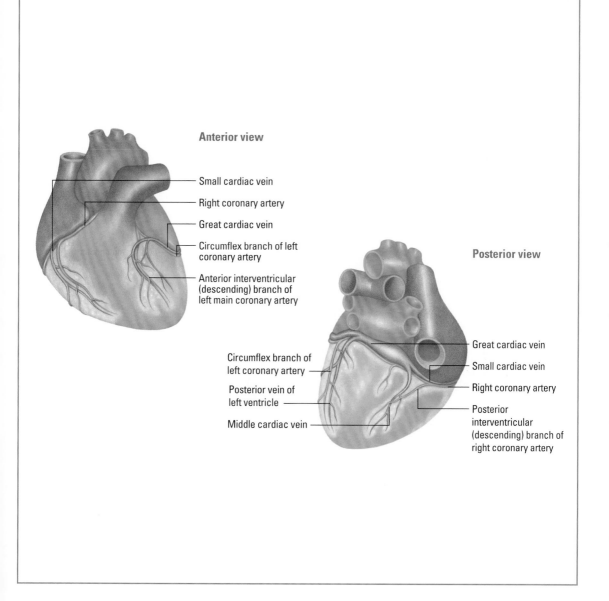

Anterior view

Small cardiac vein

Right coronary artery

Great cardiac vein

Circumflex branch of left coronary artery

Anterior interventricular (descending) branch of left main coronary artery

Posterior view

Great cardiac vein

Small cardiac vein

Right coronary artery

Posterior interventricular (descending) branch of right coronary artery

Circumflex branch of left coronary artery

Posterior vein of left ventricle

Middle cardiac vein

Organs and tissues of the immune system

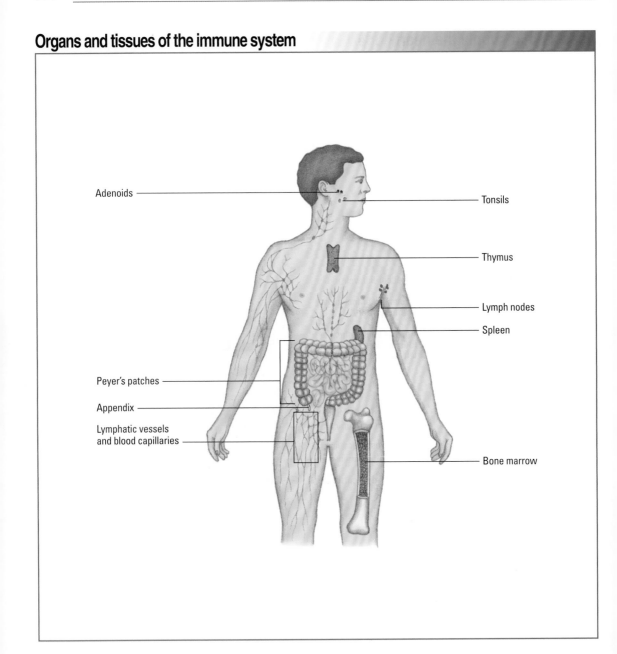

Adenoids

Tonsils

Thymus

Lymph nodes

Spleen

Peyer's patches

Appendix

Lymphatic vessels
and blood capillaries

Bone marrow

Lymphatic vessels and lymph nodes

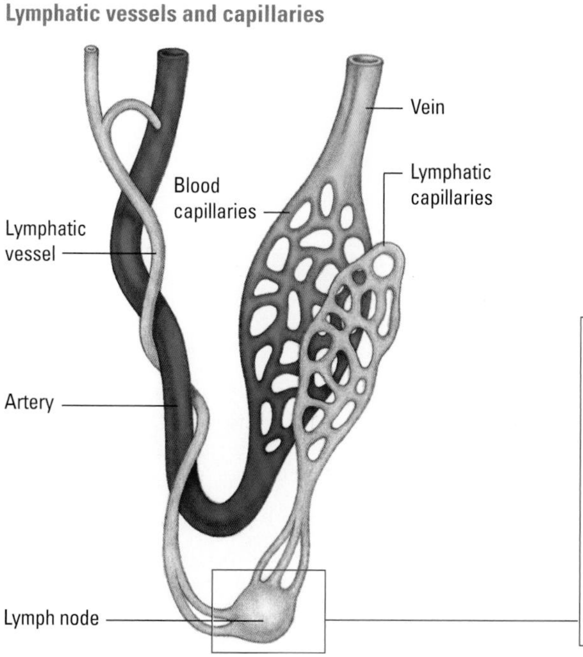

Lymphatic vessels and capillaries

Vein

Blood
capillaries

Lymphatic
capillaries

Lymphatic
vessel

Artery

Lymph node

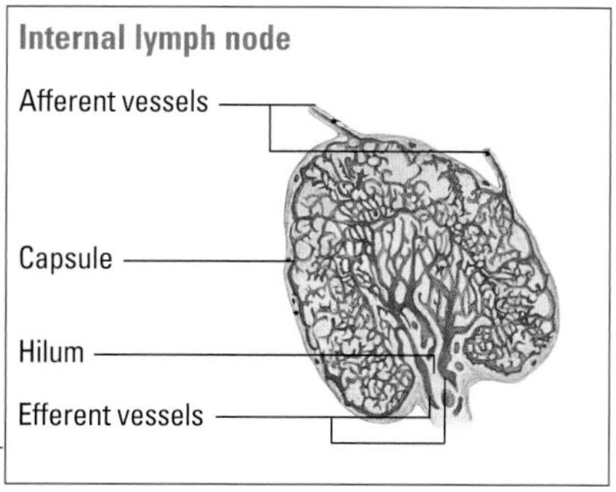

Internal lymph node

Afferent vessels

Capsule

Hilum

Efferent vessels

Structures of the respiratory system

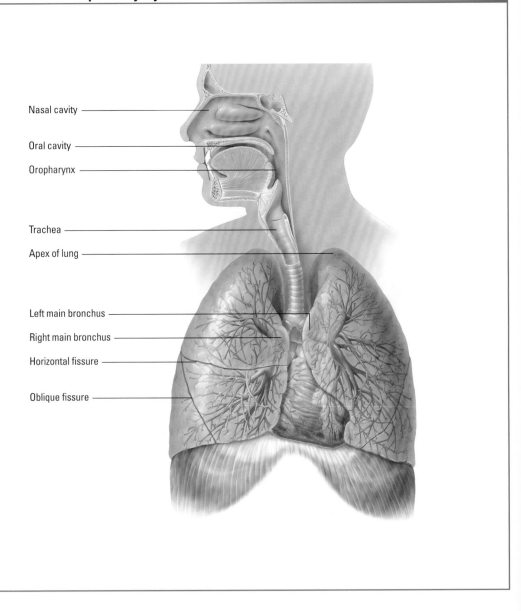

Nasal cavity

Oral cavity

Oropharynx

Trachea

Apex of lung

Left main bronchus

Right main bronchus

Horizontal fissure

Oblique fissure

A close look at a pulmonary airway

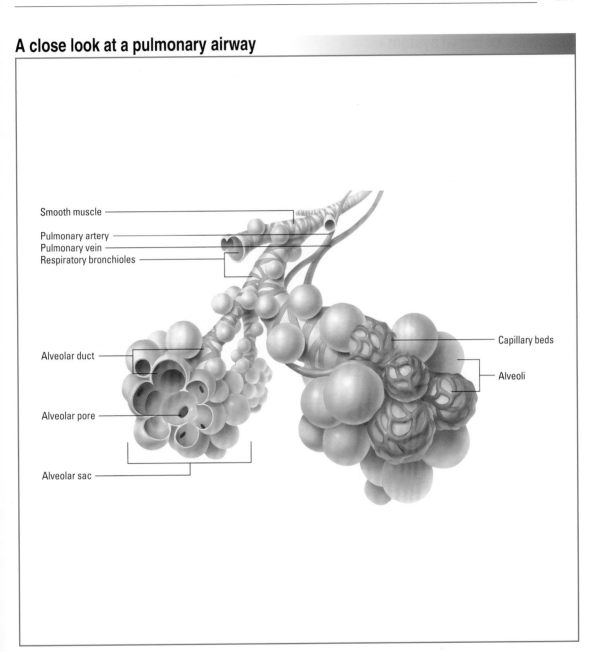

Smooth muscle

Pulmonary artery

Pulmonary vein

Respiratory bronchioles

Alveolar duct

Alveolar pore

Alveolar sac

Capillary beds

Alveoli

Structures of the GI system

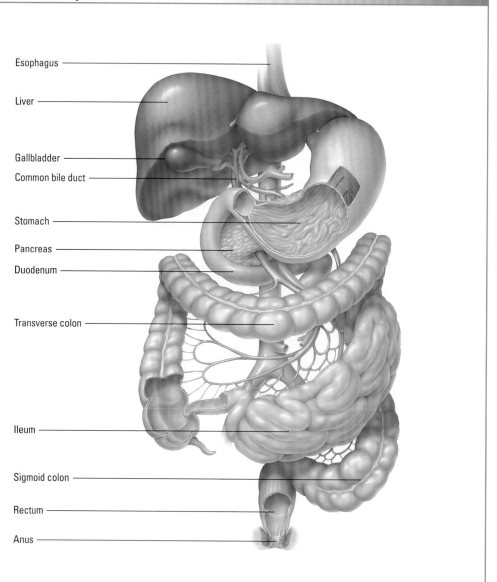

Esophagus

Liver

Gallbladder

Common bile duct

Stomach

Pancreas

Duodenum

Transverse colon

Ileum

Sigmoid colon

Rectum

Anus

Oral cavity

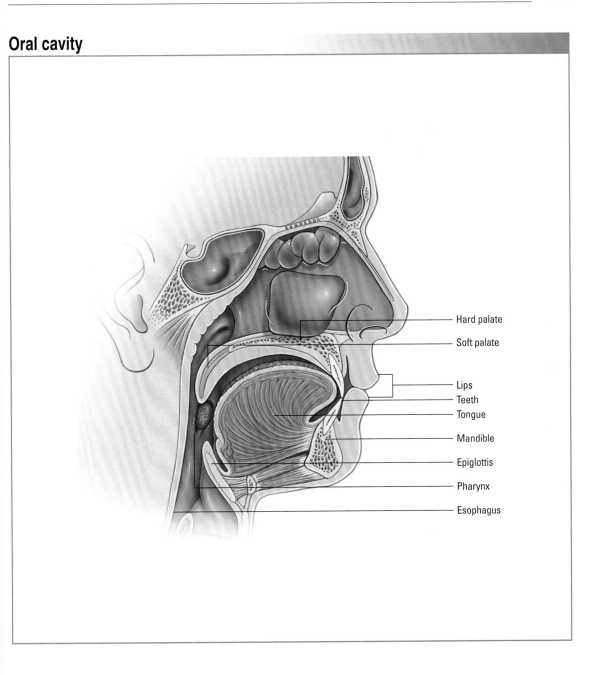

Hard palate

Soft palate

Lips

Teeth

Tongue

Mandible

Epiglottis

Pharynx

Esophagus

A close look at the urinary system

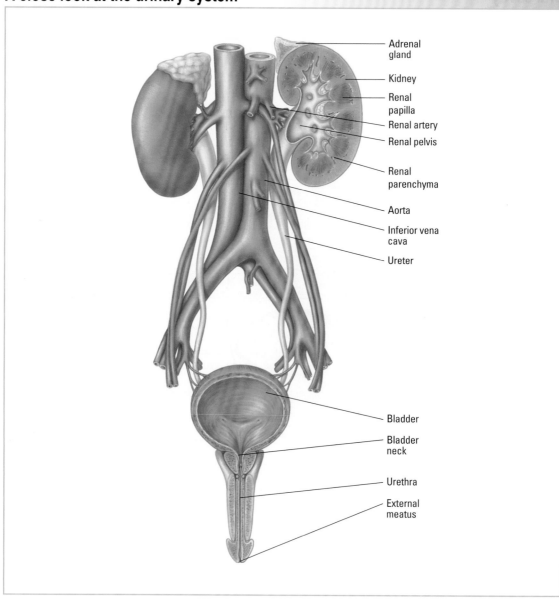

- Adrenal gland
- Kidney
- Renal papilla
- Renal artery
- Renal pelvis
- Renal parenchyma
- Aorta
- Inferior vena cava
- Ureter
- Bladder
- Bladder neck
- Urethra
- External meatus

Structure of the nephron

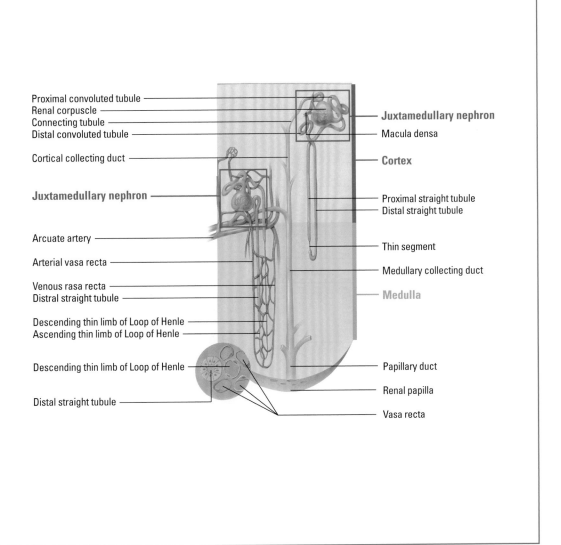

Proximal convoluted tubule

Renal corpuscle

Connecting tubule

Distal convoluted tubule

Cortical collecting duct

Juxtamedullary nephron

Arcuate artery

Arterial vasa recta

Venous rasa recta

Distral straight tubule

Descending thin limb of Loop of Henle

Ascending thin limb of Loop of Henle

Descending thin limb of Loop of Henle

Distal straight tubule

Juxtamedullary nephron

Macula densa

Cortex

Proximal straight tubule

Distal straight tubule

Thin segment

Medullary collecting duct

Medulla

Papillary duct

Renal papilla

Vasa recta

Structures of the male reproductive system

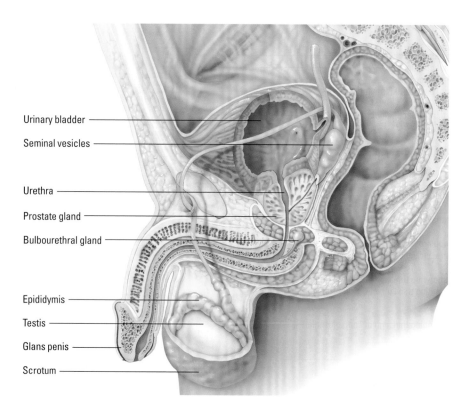

Urinary bladder

Seminal vesicles

Urethra

Prostate gland

Bulbourethral gland

Epididymis

Testis

Glans penis

Scrotum

Female external genitalia

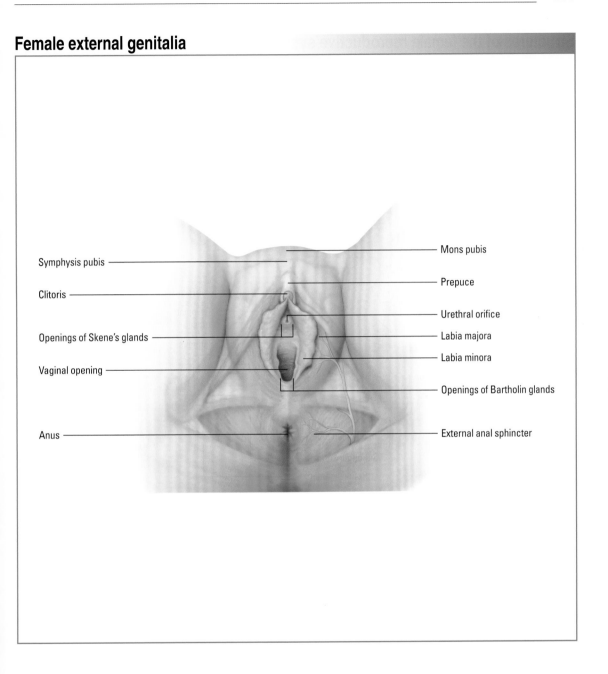

Symphysis pubis

Clitoris

Openings of Skene's glands

Vaginal opening

Anus

Mons pubis

Prepuce

Urethral orifice

Labia majora

Labia minora

Openings of Bartholin glands

External anal sphincter

Structures of the female reproductive system

Lateral view

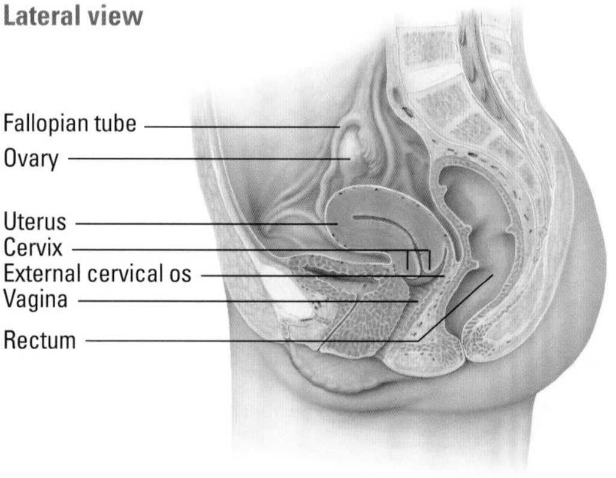

Fallopian tube

Ovary

Uterus
Cervix
External cervical os
Vagina

Rectum

Anterior cross-sectional view

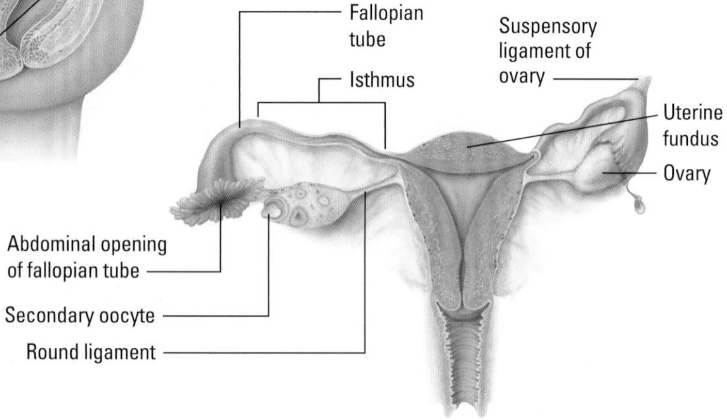

Fallopian tube

Isthmus

Suspensory ligament of ovary

Uterine fundus

Ovary

Abdominal opening of fallopian tube

Secondary oocyte

Round ligament

The female breast

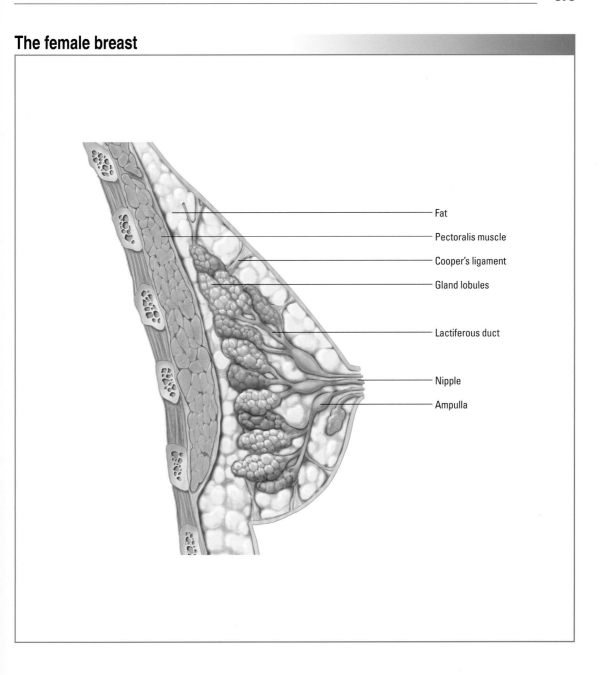

Fat

Pectoralis muscle

Cooper's ligament

Gland lobules

Lactiferous duct

Nipple

Ampulla

Index

Note: Page numbers followed by `f` denote figures those followed by `t` denote tables.

377

Note: Page numbers followed by `*f*' denote figures those followed by `*t*' denote tables.

Note: Page numbers followed by '*f*' denote figures those followed by '*t*' denote tables.

Note: Page numbers followed by `*f*' denote figures those followed by `*t*' denote tables.

Note: Page numbers followed by `f` denote figures those followed by `t` denote tables.

Note: Page numbers followed by `f' denote figures those followed by `t' denote tables.

Note: Page numbers followed by `*f*' denote figures those followed by `*t*' denote tables.

Note: Page numbers followed by `f` denote figures those followed by `t` denote tables.